STUDENT SOLUTIONS MANUAL

CHAPTERS 1–19

PHYSICS

FOR SCIENTISTS AND ENGINEERS THIRD EDITION

A STRATEGIC APPROACH

Randall D. Knight

Larry Smith
Snow College

Brett Kraabel
PhD-Physics, University of Santa Barbara

PEARSON

Boston Columbus Indianapolis New York San Francisco Upper Saddle River
Amsterdam Cape Town Dubai London Madrid Milan Munich Paris Montréal Toronto
Delhi Mexico City São Paulo Sydney Hong Kong Seoul Singapore Taipei Tokyo

Publisher: James Smith

Senior Development Editor: Alice Houston, Ph.D.

Senior Project Editor: Martha Steele

Assistant Editor: Peter Alston

Media Producer: Kelly Reed

Senior Administrative Assistant: Cathy Glenn

Director of Marketing: Christy Lesko

Executive Marketing Manager: Kerry McGinnis

Managing Editor: Corinne Benson

Production Project Manager: Beth Collins

Manufacturing Buyer: Jeffrey Sargent

Production Management, Illustration, and Composition: PreMediaGlobal, Inc.

Cover Design: Riezebos-Holzbaur Design Group and Seventeenth Street Studios

Cover Photo Illustration: Riezebos-Holzbaur Design Group

Printer: Edwards Brothers Malloy

ISBN 13: 978-0-321-74767-9

ISBN 10: 0-321-74767-4

4 5 6 7 8 9 10—EBM—17 16 15

www.pearsonhighered.com

Contents

Preface v

PART I   **Newton's Laws**

Chapter 1   Concepts of Motion 1-1
                Exercises 1.3, 1.5, 1.7, 1.9, 1.13, 1.15, 1.21, 1.23, 1.25, 1.29, 1.33
                Problems 1.35, 1.39, 1.41, 1.45, 1.47, 1.49, 1.51, 1.55

Chapter 2   Kinematics in One Dimension 2-1
                Exercises 2.1, 2.5, 2.9, 2.11, 2.15, 2.17, 2.19, 2.23, 2.25
                Problems 2.27, 2.31, 2.33, 2.37, 2.39, 2.45, 2.47, 2.49, 2.51, 2.57, 2.59, 2.63,
                2.65, 2.67, 2.71

Chapter 3   Vectors and Coordinate Systems 3-1
                Exercises 3.1, 3.3, 3.7, 3.11, 3.15, 3.17
                Problems 3.19, 3.23, 3.27, 3.29, 3.31, 3.35, 3.37, 3.39, 3.43

Chapter 4   Kinematics in Two Dimensions 4-1
                Exercises 4.3, 4.5, 4.9, 4.11, 4.13, 4.15, 4.21, 4.23, 4.25, 4.29, 4.31, 4.33
                Problems 4.39, 4.41, 4.45, 4.47, 4.49, 4.51, 4.55, 4.59, 4.61, 4.65, 4.67, 4.71, 4.73

Chapter 5   Force and Motion 5-1
                Exercises 5.1, 5.3, 5.9, 5.11, 5.13, 5.17, 5.21, 5.23, 5.25
                Problems 5.29, 5.31, 5.35, 5.37, 5.39, 5.43, 5.47, 5.49, 5.51

Chapter 6   Dynamics I: Motion Along a Line 6-1
                Exercises 6.1, 6.3, 6.7, 6.9, 6.13, 6.15, 6.19, 6.25
                Problems 6.29, 6.31, 6.33, 6.37, 6.41, 6.43, 6.47, 6.49, 6.53, 6.55, 6.57, 6.63, 6.65

Chapter 7   Newton's Third Law 7-1
                Exercises 7.1, 7.5, 7.7, 7.11, 7.13, 7.17
                Problems 7.21, 7.23, 7.27, 7.29, 7.31, 7.35, 7.37, 7.41, 7.45, 7.47, 7.49

Chapter 8   Dynamics II: Motion in a Plane 8-1
                Exercises 8.1, 8.3, 8.7, 8.9, 8.13, 8.15, 8.19
                Problems 8.21, 8.27, 8.29, 8.31, 8.35, 8.37, 8.39, 8.43, 8.47, 8.51, 8.53, 8.55

PART II   **Conservation Laws**

Chapter 9   Impulse and Momentum 9-1
                Exercises 9.1, 9.5, 9.7, 9.9, 9.13, 9.17, 9.19, 9.21, 9.25
                Problems 9.29, 9.31, 9.33, 9.39, 9.41, 9.43, 9.45, 9.51, 9.53, 9.55, 9.57, 9.63, 9.65

Chapter 10 Energy 10-1
                Exercises 10.3, 10.5, 10.9, 10.11, 10.13, 10.15, 10.19, 10.21, 10.25, 10.27, 10.31
                Problems 10.33, 10.37, 10.39, 10.43, 10.45, 10.49, 10.51, 10.53, 10.57, 10.59,
                10.61

Chapter 11   Work 11-1

           Exercises 11.3, 11.5, 11.7, 11.11, 11.13, 11.15, 11.21, 11.23, 11.27, 11.29,
                 11.31, 11.33

           Problems 11.39, 11.41, 11.45, 11.47, 11.49, 11.53, 11.55, 11.57, 11.61, 11.65,
                 11.67

PART III   **Applications of Newtonian Mechanics**

Chapter 12   Rotation of a Rigid Body 12-1

           Exercises 12.1, 12.5, 12.9, 12.11, 12.13, 12.17, 12.19, 12.23, 12.25, 12.27, 12.31,
                 12.35, 12.39, 12.41, 12.43, 12.47

           Problems 12.51, 12.53, 12.57, 12.59, 12.63, 12.65, 12.67, 12.71, 12.73, 12.75, 12.81

Chapter 13   Newton's Theory of Gravity 13-1

           Exercises 13.1, 13.5, 13.9, 13.11, 13.13, 13.17, 13.19, 13.23

           Problems 13.27, 13.29, 13.33, 13.35, 13.41, 13.45, 13.47, 13.49, 13.51, 13.57

Chapter 14   Oscillations 14-1

           Exercises 14.3, 14.5, 14.7, 14.9, 14.13, 14.17, 14.19, 14.21, 14.25, 14.29

           Problems 14.33, 14.35, 14.37, 14.39, 14.43, 14.47, 14.51, 14.53, 14.57,
                 14.59, 14.61, 14.63, 14.67, 14.69, 14.73

Chapter 15   Fluids and Elasticity 15-1

           Exercises 15.1, 15.3, 15.9, 15.11, 15.13, 15.17, 15.19, 15.21, 15.25, 15.27

           Problems 15.31, 15.35, 15.37, 15.39, 15.43, 15.47, 15.51, 15.53, 15.55,
                 15.57, 15.61, 15.63, 15.67, 15.69

PART IV   **Thermodynamics**

Chapter 16   A Macroscopic Description of Matter 16-1

           Exercises 16.1, 16.3, 16.9, 16.11, 16.13, 16.15, 16.19, 16.21, 16.25, 16.27

           Problems 16.33, 16.35, 16.39, 16.41, 16.45, 16.47, 16.51, 16.53, 16.55, 16.57,
                 16.61, 16.63

Chapter 17   Work, Heat, and the First Law of Thermodynamics 17-1

           Exercises 17.1, 17.5, 17.9, 17.11, 17.13, 17.17, 17.21, 17.23, 17.25, 17.29

           Problems 17.33, 17.35, 17.37, 17.39, 17.43, 17.47, 17.49, 17.53, 17.55,
                 17.57, 17.61, 17.63, 17.67, 17.69

Chapter 18   The Micro/Macro Connection 18-1

           Exercises 18.1, 18.5, 18.7, 18.9, 18.13, 18.17, 18.19, 18.23, 18.25, 18.29,
                 18.33, 18.35

           Problems 18.39, 18.41, 18.43, 18.45, 18.49, 18.53, 18.55, 18.59, 18.61

Chapter 19   Heat Engines and Refrigerators 19-1

           Exercises 19.1, 19.5, 19.9, 19.11, 19.15, 19.17, 19.19, 19.21, 19.25, 19.29

           Problems 19.33, 19.35, 19.37, 19.41, 19.45, 19.47, 19.51, 19.53, 19.57,
                 19.59, 19.61

Preface

This *student Solutions Manual* is intended to provide you with examples of good problem-solving techniques and strategies. To achieve that, the solutions presented here attempt to:

- Follow, in detail, the problem-solving strategies presented in the text.
- Articulate the reasoning that must be done before computation.
- Illustrate how to use drawings effectively.
- Demonstrate how to utilize graphs, ratios, units, and the many other "tactics" that must be successfully mastered and marshaled if a problem-solving strategy is to be effective.
- Show examples of assessing the reasonableness of a solution.
- Comment on the significance of a solution or on its relationship to other problems.

We recommend you try to solve each problem on your own before you read the solution. Simply reading solutions, without first struggling with the issues, has limited educational value.

As you work through each solution, make sure you understand how and why each step is taken. See if you can understand which aspects of the problem made this solution strategy appropriate. You will be successful on exams not by memorizing solutions to particular problems but by coming to recognize which kinds of problem-solving strategies go with which types of problems.

We have made every effort to be accurate and correct in these solutions. However, if you do find errors or ambiguities, we would be very grateful to hear from you. Please have your instructor contact the Pearson Education sales representative.

Acknowledgments for the First Edition

We are grateful for many helpful comments from Susan Cable, Randall Knight, and Steve Stonebraker. We express appreciation to Susan Emerson, who typed the word-processing manuscript, for her diligence in interpreting our handwritten copy. Finally, we would like to acknowledge the support from the Addison Wesley staff in getting the work into a publishable state. Our special thanks to Liana Allday, Alice Houston, and Sue Kimber for their willingness and preparedness in providing needed help at all times.

Pawan Kahol
Missouri State University

Donald Foster
Wichita State University

Acknowledgments for the Second Edition

I would like to acknowledge the patient support of my wife, Holly, who knows what is important.

Larry Smith
Snow College

I would like to acknowledge the assistance and support of my wife, Alice Nutter, who helped type many problems and was patient while I worked weekends.

Scott Nutter
Northern Kentucky University

Acknowledgments for the Third Edition

To Holly, Ryan, Timothy, Nathan, Tessa, and Tyler, who make it all worthwhile.

Larry Smith
Snow College

I gratefully acknowledge the assistance of the staff at Physical Sciences Communication.

Brett Kraabel
PhD-University of Santa Barbara

CONCEPTS OF MOTION

Exercises and Problems

Section 1.1   Motion Diagrams

1.3.   Model: We will assume that the term "quickly" used in the problem statement means a time that is short compared to 30 s.
Solve:

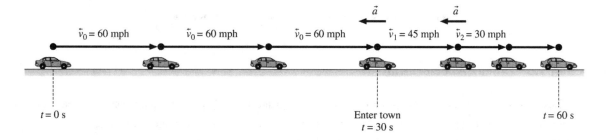

Assess: Notice that the acceleration vector points in the direction opposite to the velocity vector because the car is decelerating.

Section 1.3   Position and Time

Section 1.4   Velocity

1.5.   Model: We model the ball's motion from the instant after it is released, when it has zero velocity, to the instant before it hits the ground, when it will have its maximum velocity.

Solve:

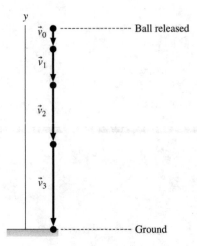

Assess: Notice that the "particle" we have drawn has a finite dimensions, so it appears as if the bottom half of this "particle" has penetrated into the ground in the bottom frame. This is not really the case; our mental particle has no size and is located at the tip of the velocity vector arrow.

1.7.   Solve: The player starts with an initial velocity but as he slides he moves slower and slower until coming to rest.

Section 1.5 Linear Acceleration

1.9.   Solve: (a) Let $\vec{v}_0$ be the velocity vector between points 0 and 1 and $\vec{v}_1$ be the velocity vector between points 1 and 2. Speed v_1 is greater than speed v_0 because more distance is covered in the same interval of time.
(b) Acceleration is found by the method of Tactics Box 1.3.

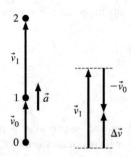

Assess: The acceleration vector points in the same direction as the velocity vectors, which makes sense because the speed is increasing.

1.13.   Model: Represent the (child + sled) system as a particle.

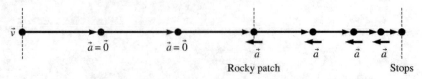

Visualize: The dots in the figure are equally spaced until the sled encounters a rocky patch. Equidistant dots on a single line indicate constant average velocity. On encountering a rocky patch, the average velocity decreases and the sled comes to a stop. This part of the motion is indicated by a decreasing separation between the dots.

1.15. **Model:** Represent the tile as a particle.
Visualize: Starting from rest, the tile's velocity increases until it hits the water surface. This part of the motion is represented by dots with increasing separation, indicating increasing average velocity. After the tile enters the water, it settles to the bottom at roughly constant speed, so this part of the motion is represented by equally spaced dots.

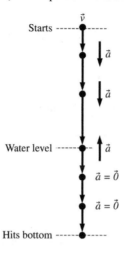

Section 1.7 Solving Problems in Physics

1.21. **Visualize:** The bicycle move forward with an acceleration of 1.5 m/s². Thus, the velocity will increase by 1.5 m/s each second of motion.

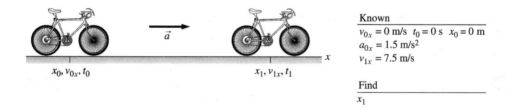

Known
$v_{0x} = 0$ m/s $t_0 = 0$ s $x_0 = 0$ m
$a_{0x} = 1.5$ m/s²
$v_{1x} = 7.5$ m/s

Find
x_1

Section 1.8 Units and Significant Figures

1.23. **Solve:** **(a)** $6.15 \text{ ms} = (6.15 \text{ ms})\left(\dfrac{1 \text{ s}}{10^3 \text{ ms}}\right) = 6.15 \times 10^{-3} \text{ s}$

(b) $27.2 \text{ km} = (27.2 \text{ km})\left(\dfrac{10^3 \text{ m}}{1 \text{ km}}\right) = 27.2 \times 10^3 \text{ m}$

(c) $112 \text{ km/hour} = \left(112 \dfrac{\text{km}}{\text{hour}}\right)\left(\dfrac{10^3 \text{ m}}{1 \text{ km}}\right)\left(\dfrac{1 \text{ hour}}{3600 \text{ s}}\right) = 31.1 \text{ m/s}$

(d) $72 \ \mu m/ms = \left(72 \dfrac{\mu m}{ms} \right) \left(\dfrac{1 \ m}{10^6 \ \mu m} \right) \left(\dfrac{10^3 \ ms}{1 \ s} \right) = 7.2 \times 10^{-2} \ m/s$

1.25. **Solve:** **(a)** $3 \ hour = (3 \ hour) \left(\dfrac{3600 \ s}{1 \ hour} \right) = 10,800 \ s \approx 1 \times 10^4 \ s$

(b) $2 \ day = (2 \ day) \left(\dfrac{24 \ hours}{1 \ day} \right) \left(\dfrac{3600 \ s}{1 \ hour} \right) = 1.73 \times 10^5 \ s \approx 2 \times 10^5 \ s$

(c) $1 \ year = (1 \ year) \left(\dfrac{365.25 \ days}{1 \ year} \right) \left(\dfrac{8.64 \times 10^4 \ s}{1 \ day} \right) = 3.16 \times 10^7 \ s \approx 3 \times 10^7 \ s$

(d) $215 \ ft/s = \left(215 \dfrac{ft}{s} \right) \left(\dfrac{12 \ inch}{1 \ ft} \right) \left(\dfrac{1 \ m}{39.37 \ inch} \right) = 65.5 \ m/s$

Assess: The results are given to appropriate number of significant figures.

1.29. **Solve:** **(a)** $(12.5)^3 = 1.95 \times 10^3$.

(b) $12.5 \times 5.21 = 65.1$

(c) $\sqrt{12.5} - 1.2 = 3.54 - 1.2 = 2.3$

(d) $12.5^{-1} = 1.00/12.5 = 0.0800 = 8.00 \times 10^{-2}$

1.33. **Solve:** My barber trims about an inch of hair when I visit him every month for a haircut. The rate of hair growth is

$$\left(\dfrac{1 \ inch}{1 \ month} \right) \left(\dfrac{2.54 \ cm}{1 \ inch} \right) \left(\dfrac{10^{-2} \ m}{1 \ cm} \right) \left(\dfrac{1 \ month}{30 \ days} \right) \left(\dfrac{1 \ day}{24 \ h} \right) \left(\dfrac{1 \ h}{3600 \ s} \right) = 9.8 \times 10^{-9} \ m/s$$

$$= \left(9.8 \times 10^{-9} \dfrac{m}{s} \right) \left(\dfrac{10^6 \ \mu m}{1 \ m} \right) \left(\dfrac{3600 \ s}{1 \ h} \right) \approx 40 \ \mu m/h$$

1.35. **Model:** Represent the jet as a particle for the motion diagram.
Visualize:

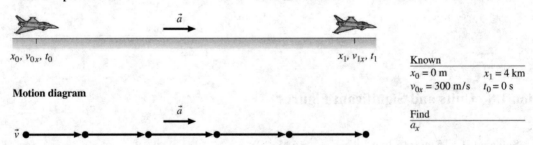

1.39. **Model:** Represent Santa Claus as a particle for the motion diagram.

Visualize:

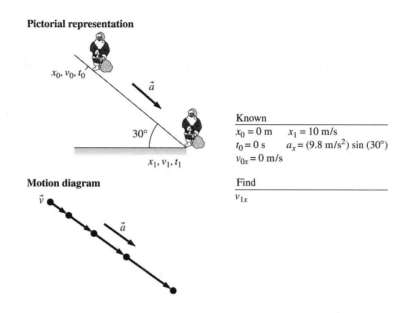

1.41.   Model: Represent the car as a particle for the motion diagram.
Visualize:

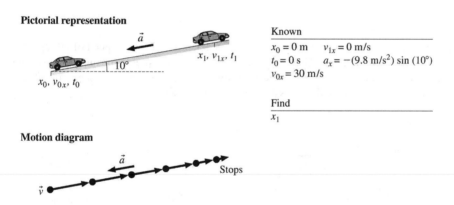

1.45.   Solve: A car coasts along at 30 m/s and arrives at a hill. The car decelerates as it coasts up the hill. At the top, the road levels and the car continues coasting along the road at a reduced speed.

1.47.   Solve: A ball is dropped from a height to check its rebound properties. It rebounds to 80% of its original height.

1.49.   Solve:
(a)

(b) A train moving at 100 km/hour slows down in 10 s to a speed of 60 km/hour as it enters a tunnel. The driver maintains this constant speed for the entire length of the tunnel that takes the train a time of 20 s to traverse. Find the length of the tunnel.

(c)

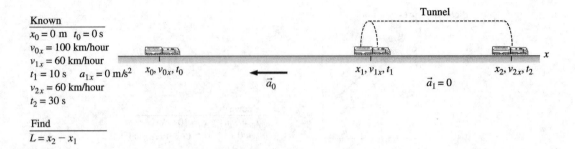

Known
$x_0 = 0$ m $t_0 = 0$ s
$v_{0x} = 100$ km/hour
$v_{1x} = 60$ km/hour
$t_1 = 10$ s $a_{1x} = 0$ m/s^2
$v_{2x} = 60$ km/hour
$t_2 = 30$ s

Find
$L = x_2 - x_1$

1.51. Solve:
(a)

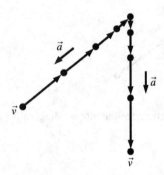

(b) Jeremy has perfected the art of steady acceleration and deceleration. From a speed of 60 mph he brakes his car to rest in 10 s with a constant deceleration. Then he turns into an adjoining street. Starting from rest, Jeremy accelerates with exactly the same magnitude as his earlier deceleration and reaches the same speed of 60 mph over the same distance in exactly the same time. Find the car's acceleration or deceleration.

(c)

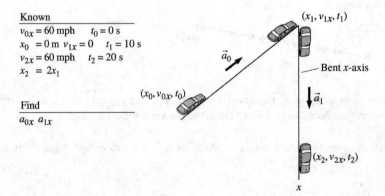

Known
$v_{0x} = 60$ mph $t_0 = 0$ s
$x_0 = 0$ m $v_{1x} = 0$ $t_1 = 10$ s
$v_{2x} = 60$ mph $t_2 = 20$ s
$x_2 = 2x_1$

Find
a_{0x} a_{1x}

1.55. **Model:** In the particle model, the car is represented as a dot.
Solve:

(a)

Time t (s)	Position x (m)
0	1200
10	975
20	825
30	750
40	700
50	650
60	600
70	500
80	300
90	0

(b)

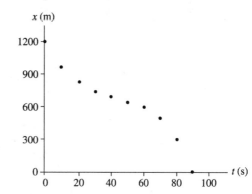

2

KINEMATICS IN ONE DIMENSION

Exercises and Problems

Section 2.1 Uniform Motion

2.1. Model: Cars will be treated by the particle model.

Visualize:

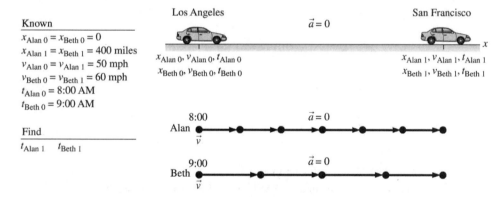

Pictorial representation

Known

$x_{\text{Alan 0}} = x_{\text{Beth 0}} = 0$
$x_{\text{Alan 1}} = x_{\text{Beth 1}} = 400$ miles
$v_{\text{Alan 0}} = v_{\text{Alan 1}} = 50$ mph
$v_{\text{Beth 0}} = v_{\text{Beth 1}} = 60$ mph
$t_{\text{Alan 0}} = 8{:}00$ AM
$t_{\text{Beth 0}} = 9{:}00$ AM

Find

$t_{\text{Alan 1}}$ $t_{\text{Beth 1}}$

Solve: Beth and Alan are moving at a constant speed, so we can calculate the time of arrival as follows:

$$v = \frac{\Delta x}{\Delta t} = \frac{x_1 - x_0}{t_1 - t_0} \Rightarrow t_1 = t_0 + \frac{x_1 - x_0}{v}$$

Using the known values identified in the pictorial representation, we find:

$$t_{\text{Alan 1}} = t_{\text{Alan 0}} + \frac{x_{\text{Alan 1}} - x_{\text{Alan 0}}}{v} = 8{:}00 \text{ AM} + \frac{400 \text{ mile}}{50 \text{ miles/hour}} = 8{:}00 \text{ AM} + 8 \text{ hr} = 4{:}00 \text{ PM}$$

$$t_{\text{Beth 1}} = t_{\text{Beth 0}} + \frac{x_{\text{Beth 1}} - x_{\text{Beth 0}}}{v} = 9{:}00 \text{ AM} + \frac{400 \text{ mile}}{60 \text{ miles/hour}} = 9{:}00 \text{ AM} + 6.67 \text{ hr} = 3{:}40 \text{ PM}$$

(a) Beth arrives first.
(b) Beth has to wait $t_{\text{Alan 1}} - t_{\text{Beth 1}} = 20$ minutes for Alan.

Assess: Times of the order of 7 or 8 hours are reasonable in the present problem.

Section 2.2 Instantaneous Velocity

Section 2.3 Finding Position from Velocity

2.5. Solve: (a) We can obtain the values for the velocity-versus-time graph from the equation $v = \Delta s/\Delta t$.

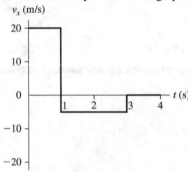

(b) There is only one turning point. At $t = 1$ s the velocity changes from $+20$ m/s to -5 m/s, thus reversing the direction of motion. At $t = 3$ s, there is an abrupt change in motion from -5 m/s to rest, but there is no reversal in motion.

Section 2.4 Motion with Constant Acceleration

2.9. Visualize: The object has a constant velocity for 2 s and then speeds up between $t = 2$ and $t = 4$.
Solve: A constant velocity from $t = 0$ s to $t = 2$ s means zero acceleration. On the other hand, a linear increase in velocity between $t = 2$ s and $t = 4$ s implies a constant positive acceleration which is the slope of the velocity line.

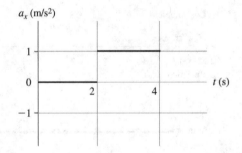

2.11. Solve: (a) At $t = 2.0$ s, the position of the particle is

$$x_{2\,s} = 2.0 \text{ m} + \text{area under velocity graph from } t = 0 \text{ s to } t = 2.0 \text{ s}$$

$$= 2.0 \text{ m} + \frac{1}{2}(4 \text{ m/s})(2.0 \text{ s}) = 6 \text{ m}$$

(b) From the graph itself at $t = 2.0$ s, $v = 4$ m/s.
(c) The acceleration is

$$a_x = \frac{\Delta v_x}{\Delta t} = \frac{v_{fx} - v_{ix}}{\Delta t} = \frac{6 \text{ m/s} - 0 \text{ m/s}}{3 \text{ s}} = 2 \text{ m/s}^2$$

2.15. Model: We are using the particle model for the skater and the kinematics model of motion under constant acceleration.
Solve: Since we don't know the time of acceleration we will use

$$v_f^2 = v_i^2 + 2a(x_f - x_i)$$

$$\Rightarrow a = \frac{v_f^2 - v_i^2}{2(x_f - x_i)} = \frac{(6.0 \text{ m/s})^2 - (8.0 \text{ m/s})^2}{2(5.0 \text{ m})} = -2.8 \text{ m/s}^2$$

Assess: A deceleration of 2.8 m/s^2 is reasonable.

Section 2.5 Free Fall

2.17. Model: Represent the spherical drop of molten metal as a particle.
Visualize:

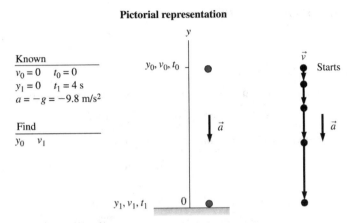

Pictorial representation

Known
$v_0 = 0$ $t_0 = 0$
$y_1 = 0$ $t_1 = 4$ s
$a = -g = -9.8$ m/s^2

Find
y_0 v_1

y_0, v_0, t_0

y_1, v_1, t_1

Solve: (a) The shot is in free fall, so we can use free fall kinematics with $a = -g$. The height must be such that the shot takes 4 s to fall, so we choose $t_1 = 4$ s. Then,

$$y_1 = y_0 + v_0(t_1 - t_0) - \frac{1}{2}g(t_1 - t_0)^2 \Rightarrow y_0 = \frac{1}{2}gt_1^2 = \frac{1}{2}(9.8 \ \text{m/s}^2)(4 \ \text{s})^2 = 78.4 \ \text{m}$$

(b) The impact velocity is $v_1 = v_0 - g(t_1 - t_0) = -gt_1 = -39.2$ m/s.

Assess: Note the minus sign. The question asked for *velocity*, not speed, and the y-component of $\vec{v}$ is negative because the vector points downward.

2.19. Model: We model the ball as a particle.
Visualize:

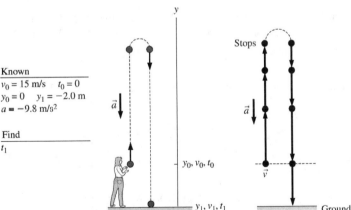

Pictorial representation

Known
$v_0 = 15$ m/s $t_0 = 0$
$y_0 = 0$ $y_1 = -2.0$ m
$a = -9.8$ m/s^2

Find
t_1

Stops

y_0, v_0, t_0

y_1, v_1, t_1

Ground

Solve: Once the ball leaves the student's hand, the ball is in free fall and its acceleration is equal to the free-fall acceleration g that always acts vertically downward toward the center of the earth. According to the constant-acceleration kinematic equations of motion

$$y_1 = y_0 + v_0 \Delta t + \frac{1}{2}a\Delta t^2$$

Substituting the known values

$$-2 \ \text{m} = 0 \ \text{m} + (15 \ \text{m/s})t_1 + (1/2)(-9.8 \ \text{m/s}^2)t_1^2$$

One solution of this quadratic equation is $t_1 = 3.2$ s. The other root of this equation yields a negative value for t_1, which is not valid for this problem.

Assess: A time of 3.2 s is reasonable.

Section 2.7 Instantaneous Acceleration

2.23. Solve: $x = (2t^2 - t + 1)$ m

(a) The position $t = 2$ s is $x_{2s} = [2(2)^2 - 2 + 1]$ m $= 7$ m.

(b) The velocity is the derivative $v = dx/dt$ and the velocity at $t = 2$ s is calculated as follows:

$$v = (4t - 1) \text{ m/s} \Rightarrow v_{2s} = [4(2) - 1] \text{ m/s} = 7 \text{ m/s}$$

(c) The acceleration is the derivative $a = dv/dt$ and the acceleration at $t = 2$ s is calculated as follows:

$$a = (4) \text{ m/s}^2 \Rightarrow a_{2s} = 4 \text{ m/s}^2$$

2.25. Solve: The formula for the particle's velocity is given by

$$v_f = v_i + \text{area under the acceleration curve between } t_i \text{ and } t_f$$

For $t = 4$ s, we get

$$v_{4s} = 8 \text{ m/s} + \frac{1}{2}(4 \text{ m/s}^2)4 \text{ s} = 16 \text{ m/s}$$

Assess: The acceleration is positive but decreases as a function of time. The initial velocity of 8.0 m/s will therefore increase. A value of 16 m/s is reasonable.

2.27. Solve: The graph for particle A is a straight line from $t = 2$ s to $t = 8$ s. The slope of this line is -10 m/s, which is the velocity at $t = 7.0$ s. The negative sign indicates motion toward lower values on the x-axis. The velocity of particle B at $t = 7.0$ s can be read directly from its graph. It is -20 m/s. The velocity of particle C can be obtained from the equation

$$v_f = v_i + \text{area under the acceleration curve between } t_i \text{ and } t_f$$

This area can be calculated by adding up three sections. The area between $t = 0$ s and $t = 2$ s is 40 m/s, the area between $t = 2$ s and $t = 5$ s is 45 m/s, and the area between $t = 5$ s and $t = 7$ s is -20 m/s. We get $(10 \text{ m/s}) + (40 \text{ m/s}) + (45 \text{ m/s}) - (20 \text{ m/s}) = 75$ m/s.

2.31. Solve: (a) The velocity-versus-time graph is given by the derivative with respect to time of the position function:

$$v_x = \frac{dx}{dt} = (6t^2 - 18t) \text{ m/s}$$

For $v_x = 0$ m/s, there are two solutions to the quadratic equation: $t = 0$ s and $t = 3$ s.

(b) At the first of these solutions,

$$x(\text{at } t = 0 \text{ s}) = 2(0 \text{ s})^3 - 9(0 \text{ s})^2 + 12 = 12 \text{ m}$$

The acceleration is the derivative of the velocity function:

$$a_x = \frac{dv_x}{dt} = (12t - 18) \text{ m/s}^2 \Rightarrow a(\text{at } t = 0 \text{ s}) = -18 \text{ m/s}^2$$

At the second solution,

$$x(\text{at } t = 3 \text{ s}) = 2(3 \text{ s})^3 - 9(3 \text{ s})^2 + 12 = -15 \text{ m} \qquad a_x(\text{at } t = 3 \text{ s}) = 12(3 \text{ s}) - 18 = +18 \text{ m/s}^2$$

2.33. Solve: The position is the integral of the velocity.

$$x_1 = x_0 + \int_{t_0}^{t_1} v_x dt = x_0 + \int_0^{t_1} kt^2 dt = x_0 + \frac{1}{3}kt^3\Big|_0^{t_1} = x_0 + \frac{1}{3}kt_1^3$$

We're given that $x_0 = -9.0$ m and that the particle is at $x_1 = 9.0$ m at $t_1 = 3.0$ s. Thus

$$9.0 \text{ m} = (-9.0 \text{ m}) + \frac{1}{3}k(3.0 \text{ s})^3 = (-9.0 \text{ m}) + k(9.0 \text{ s}^3)$$

Solving for k gives $k = 2.0 \text{ m/s}^3$.

2.37. Model: Represent the ball as a particle.
Visualize: The ball moves to the right along the first track until it strikes the wall, which causes it to move to the left on a second track. The ball then descends on a third track until it reaches the fourth track, which is horizontal.
Solve:

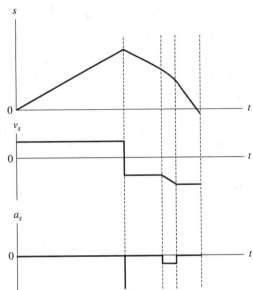

Assess: Note that the time derivative of the position graph yields the velocity graph, and the derivative of the velocity graph gives the acceleration graph.

2.39. Visualize: Please refer to Figure P2.39.
Solve:

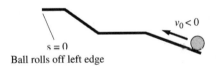

Ball rolls off left edge

2.45. Model: We will use the particle model and the constant-acceleration kinematic equations.
Visualize:

Pictorial representation

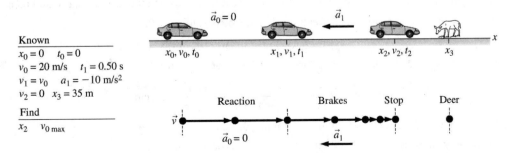

Solve: (a) To find x_2, we first need to determine x_1. Using $x_1 = x_0 + v_0(t_1 - t_0)$, we get $x_1 = 0 \text{ m} + (20 \text{ m/s})$ $(0.50 \text{ s} - 0 \text{ s}) = 10 \text{ m}$. Now,

$$v_2^2 = v_1^2 + 2a_1(x_2 - x_1) \Rightarrow 0 \text{ m}^2/\text{s}^2 = (20 \text{ m/s})^2 + 2(-10 \text{ m/s}^2)(x_2 - 10 \text{ m}) \Rightarrow x_2 = 30 \text{ m}$$

The distance between you and the deer is $(x_3 - x_2)$ or $(35 \text{ m} - 30 \text{ m}) = 5 \text{ m}$.

(b) Let us find $v_{0\,max}$ such that $v_2 = 0 \text{ m/s}$ at $x_2 = x_3 = 35 \text{ m}$. Using the following equation,

$$v_2^2 - v_{0\,max}^2 = 2a_1(x_2 - x_1) \Rightarrow 0 \text{ m}^2/\text{s}^2 - v_{0\,max}^2 = 2(-10 \text{ m/s}^2)(35 \text{ m} - x_1)$$

Also, $x_1 = x_0 + v_{0\,max}(t_1 - t_0) = v_{0\,max}(0.50 \text{ s} - 0 \text{ s}) = (0.50 \text{ s})v_{0\,max}$. Substituting this expression for x_1 in the above equation yields

$$-v_{0\,max}^2 = (-20 \text{ m/s}^2)[35 \text{ m} - (0.50 \text{ s})\,v_{0\,max}] \Rightarrow v_{0\,max}^2 + (10 \text{ m/s})v_{0\,max} - 700 \text{ m}^2/\text{s}^2 = 0$$

The solution of this quadratic equation yields $v_{0\,max} = 22 \text{ m/s}$. (The other root is negative and unphysical for the present situation.)

Assess: An increase of speed from 20 m/s to 22 m/s is very reasonable for the car to cover an additional distance of 5 m with a reaction time of 0.50 s and a deceleration of 10 m/s^2.

2.47. Model: Model the flea as a particle. Both the initial acceleration phase and the free-fall phase have constant acceleration, so use the kinematic equations.
Visualize:

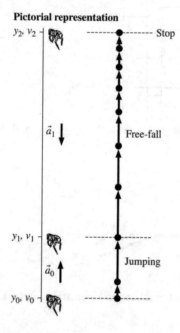

Pictorial representation

Known
$y_0 = 0 \text{ m}$ $v_0 = 0 \text{ m/s}$ $a_0 = 1000 \text{ m/s}$
$y_1 = 0.50 \text{ mm}$ $a_1 = -9.8 \text{ m/s}^2$
$v_2 = 0 \text{ m/s}$

Find
y_2

Solve: We can apply the kinematic equation $v_f^2 - v_i^2 = 2a\Delta y$ twice, once to find the take-off speed and then again to find the final height. In the first phase the acceleration is up (positive) and $v_0 = 0$.

$$v_1^2 = 2a_0(y_1 - y_0) = 2(1000 \text{ m/s}^2)(0.50 \times 10^{-3} \text{ m})v_1 = 1.0 \text{ m/s}$$

In the free fall phase the acceleration is $a_1 = -g$ and $v_1 = 1.0 \text{ m/s}$ and $v_2 = 0 \text{ m/s}$.

$$y_2 - y_1 = \frac{v_2^2 - v_1^2}{2a_1} = \frac{-v_1^2}{2(-g)} = \frac{-(1.0 \text{ m/s})^2}{2(-9.8 \text{ m/s}^2)} = 5.1 \text{ cm}$$

So the final height is $y_2 = 5.1 \text{ cm} + y_1 = 5.1 \text{ cm} + 0.50 \text{ mm} = 5.2 \text{ cm}$.

Assess: This is pretty amazing–about 10–20 times the size of a typical flea.

2.49. Model: The rocket is represented as a particle.
Visualize:

Pictorial representation

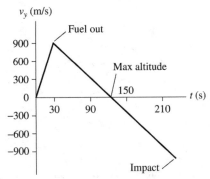

Known
$y_0 = 0$ $v_0 = 0$
$t_0 = 0$ $a_0 = 30$ m/s^2
$t_1 = 30$ s
$a_1 = -9.8$ m/s^2
$v_2 = 0$ $y_3 = 0$
$a_2 = -9.8$ m/s^2

Find
y_2 t_3

Solve: (a) There are three parts to the motion. Both the second and third parts of the motion are free fall, with $a = -g$. The maximum altitude is y_2. In the acceleration phase:

$$y_1 = y_0 + v_0(t_1 - t_0) + \frac{1}{2}a(t_1 - t_0)^2 = \frac{1}{2}at_1^2 = \frac{1}{2}(30 \text{ m/s}^2)(30 \text{ s})^2 = 13{,}500 \text{ m}$$

$$v_1 = v_0 + a(t_1 - t_0) = at_1 = (30 \text{ m/s}^2)(30 \text{ s}) = 900 \text{ m/s}$$

In the coasting phase,

$$v_2^2 = 0 = v_1^2 - 2g(y_2 - y_1) \Rightarrow y_2 = y_1 + \frac{v_1^2}{2g} = 13{,}500 \text{ m} + \frac{(900 \text{ m/s})^2}{2(9.8 \text{ m/s}^2)} = 54{,}800 \text{ m} = 54.8 \text{ km}$$

The maximum altitude is 54.8 km (≈ 33 miles).

(b) The rocket is in the air until time t_3. We already know $t_1 = 30$ s. We can find t_2 as follows:

$$v_2 = 0 \text{ m/s} = v_1 - g(t_2 - t_1) \Rightarrow t_2 = t_1 + \frac{v_1}{g} = 122 \text{ s}$$

Then t_3 is found by considering the time needed to fall 54,800 m:

$$y_3 = 0 \text{ m} = y_2 + v_2(t_3 - t_2) - \frac{1}{2}g(t_3 - t_2)^2 = y_2 - \frac{1}{2}g(t_3 - t_2)^2 \Rightarrow t_3 = t_2 + \sqrt{\frac{2y_2}{g}} = 228 \text{ s}$$

(c) The velocity increases linearly, with a slope of 30 (m/s)/s, for 30 s to a maximum speed of 900 m/s. It then begins to decrease linearly with a slope of -9.8(m/s)/s. The velocity passes through zero (the turning point at y_2) at $t_2 = 122$ s. The impact velocity at $t_3 = 228$ s is calculated to be $v_3 = v_2 - g(t_3 - t_2) = -1040$ m/s.

Assess: In reality, friction due to air resistance would prevent the rocket from reaching such high speeds as it falls, and the acceleration upward would not be constant because the mass changes as the fuel is burned, but that is a more complicated problem.

2.51. Model: We will model the lead ball as a particle and use the constant-acceleration kinematic equations.
Visualize:

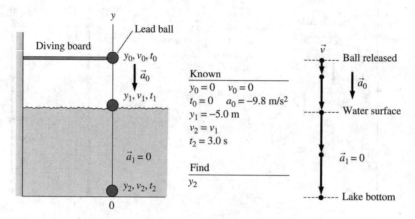

Note that the particle undergoes free fall until it hits the water surface.
Solve: The kinematics equation $y_1 = y_0 + v_0(t_1 - t_0) + \frac{1}{2}a_0(t_1 - t_0)^2$ becomes

$$-5.0 \text{ m} = 0 \text{ m} + 0 \text{ m} + \frac{1}{2}(-9.8 \text{ m/s}^2)(t_1 - 0)^2 \Rightarrow t_1 = 1.01 \text{ s}$$

Now, once again,

$$y_2 = y_1 + v_1(t_2 - t_1) + \frac{1}{2}a_1(t_2 - t_1)^2$$
$$\Rightarrow y_2 - y_1 = v_1(3.0 \text{ s} - 1.01 \text{ s}) + 0 \text{ m/s} = 1.99 \, v_1$$

v_1 is easy to determine since the time t_1 has been found. Using $v_1 = v_0 + a_0(t_1 - t_0)$, we get

$$v_1 = 0 \text{ m/s} - (9.8 \text{ m/s}^2)(1.01 \text{ s} - 0 \text{ s}) = -9.898 \text{ m/s}$$

With this value for v_1, we go back to:

$$y_2 - y_1 = 1.99 v_1 = (1.99)(-9.898 \text{ m/s}) = -19.7 \text{ m}$$

$y_2 - y_1$ is the displacement of the lead ball in the lake and thus corresponds to the depth of the lake. The negative sign shows the direction of the displacement vector.
Assess: A depth of about 60 ft for a lake is not unusual.

2.57. Model: Model the ice as a particle and use the kinematic equations for constant acceleration. Model the "very slippery block" and "smooth ramp"' as frictionless. Set the x-axis parallel to the ramp.
Visualize:

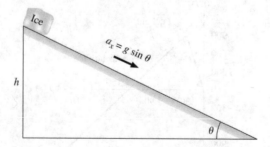

Note that the distance down the ramp is $\Delta x = h / \sin \theta$. Also $a_x = g \sin \theta$ down a frictionless ramp.

Solve:

(a) Use $v_f^2 = v_i^2 + 2a_x \Delta x$, where $v_i = 0$.

$$v_f^2 = 2a\Delta x \Rightarrow v_f = \sqrt{2(g\sin\theta)\frac{h}{\sin\theta}} = \sqrt{2gh}$$

(b) For $h = 0.30$ m,

$$v_f = \sqrt{2(9.8 \text{ m/s}^2)(0.30 \text{ m})} = 2.4 \text{ m/s}$$

This is true for both angles as the answer is independent of the angle.

Assess: We will later learn how to solve this problem in an easier way with energy.

2.59. Model: We will use the particle model and the kinematic equations at constant-acceleration.

Visualize:

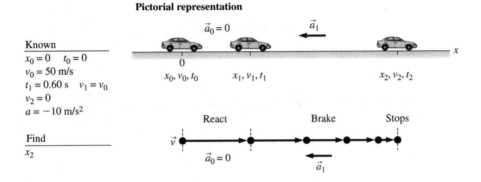

Solve: To find x_2, let us use the kinematic equation

$$v_2^2 = v_1^2 + 2a_1(x_2 - x_1) = (0 \text{ m/s})^2 = (50 \text{ m/s})^2 + 2(-10 \text{ m/s}^2)(x_2 - x_1) \Rightarrow x_2 = x_1 + 125 \text{ m}$$

Since the nail strip is at a distance of 150 m from the origin, we need to determine x_1:

$$x_1 = x_0 + v_0(t_1 - t_0) = 0 \text{ m} + (50 \text{ m/s})(0.60 \text{ s} - 0.0 \text{ s}) = 30 \text{ m}$$

Therefore, we can see that $x_2 = (30 + 125)$ m $= 155$ m. That is, he can't stop within a distance of 150 m. He is in jail.

Assess: Bob is driving at approximately 100 mph and the stopping distance is of the correct order of magnitude.

2.63. Model: The car is a particle that moves with constant linear acceleration.

Visualize:

Pictorial representation

Known
$x_0 = 0$ $v_0 = 20$ m/s
$t_0 = 0$ $t_1 = 1.0$ s
$v_1 = v_0 = 20$ m/s
$t_2 = 15$ s
$x_2 = 200$ m

Find
v_2

Solve: The reaction time is 1.0 s, and the motion during this time is

$$x_1 = x_0 + v_0(t_1 - t_0) = 0 \text{ m} + (20 \text{ m/s})(1.0 \text{ s}) = 20 \text{ m}$$

During slowing down,

$$x_2 = x_1 + v_1(t_2 - t_1) + \frac{1}{2}a_1(t_2 - t_1)^2 = 200 \text{ m}$$

$$= 20 \text{ m} + (20 \text{ m/s})(15 \text{ s} - 1.0 \text{ s}) + \frac{1}{2}a_1(15 \text{ s} - 1.0 \text{ s})^2 \Rightarrow a_1 = -1.02 \text{ m/s}^2$$

The final speed v_2 can now be obtained as

$$v_2 = v_1 + a_1(t_2 - t_1) = (20 \text{ m/s}) + (-1.02 \text{ m/s}^2)(15 \text{ s} - 1 \text{ s}) = 5.7 \text{ m/s}$$

2.65. Model: Both cars are particles that move according to the constant-acceleration kinematic equations.
Visualize:

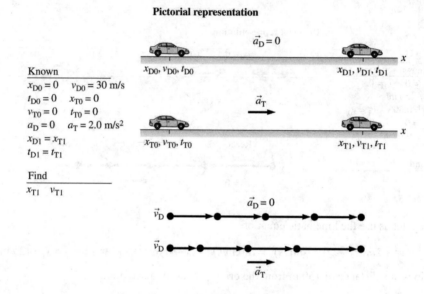

Pictorial representation

Known
$x_{D0} = 0$ $v_{D0} = 30 \text{ m/s}$
$t_{D0} = 0$ $x_{T0} = 0$
$v_{T0} = 0$ $t_{T0} = 0$
$a_D = 0$ $a_T = 2.0 \text{ m/s}^2$
$x_{D1} = x_{T1}$
$t_{D1} = t_{T1}$

Find
x_{T1} v_{T1}

Solve: (a) David's and Tina's motions are given by the following equations:

$$x_{D1} = x_{D0} + v_{D0}(t_{D1} - t_{D0}) + \frac{1}{2}a_D(t_{D1} - t_{D0})^2 = v_{D0}t_{D1}$$

$$x_{T1} = x_{T0} + v_{T0}(t_{T1} - t_{T0}) + \frac{1}{2}a_T(t_{T1} - t_{T0})^2 = 0 \text{ m} + 0 \text{ m} + \frac{1}{2}a_T t_{T1}^2$$

When Tina passes David the distances are equal and $t_{D1} = t_{T1}$, so we get

$$x_{D1} = x_{T1} \Rightarrow v_{D0}t_{D1} = \frac{1}{2}a_T t_{T1}^2 \Rightarrow v_{D0} = \frac{1}{2}a_T t_{T1} \Rightarrow t_{T1} = \frac{2v_{D0}}{a_T} = \frac{2(30 \text{ m/s})}{2.0 \text{ m/s}^2} = 30 \text{ s}$$

Using Tina's position equation,

$$x_{T1} = \frac{1}{2}a_T t_{T1}^2 = \frac{1}{2}(2.0 \text{ m/s}^2)(30 \text{ s})^2 = 900 \text{ m}$$

(b) Tina's speed v_{T1} can be obtained from

$$v_{T1} = v_{T0} + a_T(t_{T1} - t_{T0}) = (0 \text{ m/s}) + (2.0 \text{ m/s}^2)(30 \text{ s} - 0 \text{ s}) = 60 \text{ m/s}$$

Assess: This is a high speed for Tina (~134 mph) and so is David's velocity (~67 mph). Thus the large distance for Tina to catch up with David (~0.6 miles) is reasonable.

2.67. Model: Jill and the grocery cart will be treated as particles that move according to the constant-acceleration kinematic equations.
Visualize:

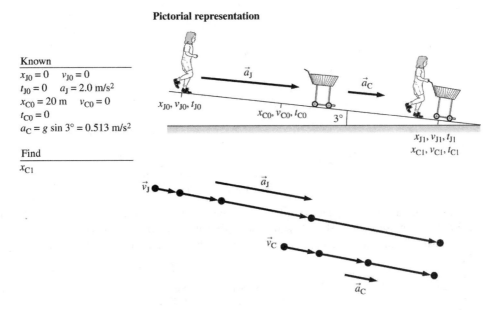

Pictorial representation

Known
$x_{J0} = 0 \quad v_{J0} = 0$
$t_{J0} = 0 \quad a_J = 2.0 \text{ m/s}^2$
$x_{C0} = 20 \text{ m} \quad v_{C0} = 0$
$t_{C0} = 0$
$a_C = g \sin 3° = 0.513 \text{ m/s}^2$

Find
x_{C1}

Solve: The final position of Jill when the cart is caught is given by

$$x_{J1} = x_{J0} + v_{J0}(t_{J1} - t_{J0}) + \frac{1}{2}a_{J0}(t_{J1} - t_{J0})^2 = 0 \text{ m} + 0 \text{ m} + \frac{1}{2}a_{J0}(t_{J1} - 0 \text{ s})^2 = \frac{1}{2}(2.0 \text{ m/s}^2)t_{J1}^2$$

The cart's position when it is caught is

$$x_{C1} = x_{C0} + v_{C0}(t_{C1} - t_{C0}) + \frac{1}{2}a_{C0}(t_{C1} - t_{C0})^2 = 20 \text{ m} + 0 \text{ m} + \frac{1}{2}(0.5 \text{ m/s}^2)(t_{C1} - 0 \text{ s})^2$$

$$= 20 \text{ m} + (0.25 \text{ m/s}^2)t_{C1}^2$$

Since $x_{J1} = x_{C1}$ and $t_{J1} = t_{C1}$, we get

$$\frac{1}{2}(2.0)t_{J1}^2 = 20 \text{ s}^2 + 0.25t_{C1}^2 \Rightarrow 0.75t_{C1}^2 = 20 \text{ s}^2 \Rightarrow t_{C1} = 5.16 \text{ s}$$

$$\Rightarrow x_{C1} = 20 \text{ m} + (0.25 \text{ m/s}^2)t_{C1}^2 = 20 \text{ m} + (0.25 \text{ m/s}^2)(5.16 \text{ s})^2 = 26.7 \text{ m}$$

So, the cart has moved 6.7 m.

2.71. Model: Model the car as a particle. Ignore air resistance. Hard braking means the wheels are locked (not turning) and the car is in full skid. For convenience, assume the car is skidding to the right.
Visualize: We use the kinematic equation $v_f^2 - v_i^2 = 2a\Delta x$. In this case $v_f = 0$ and $a < 0$, but since we want the deceleration (which is the absolute value of the acceleration) we drop the negative signs. Relabel v_i as v. We'll call the beginning of the skid mark the origin so that $x_i = 0$ and the skid length is $\Delta x = x$. Making these substitutions leaves

$$v^2 = (2a)x$$

So we expect a graph of v^2 vs. x to produce a straight line whose slope is $2a$ and whose intercept is zero. Compare to $y = mx + b$ where $y = v^2$, $m = 2a$, $x = x$, and $b = 0$.
Solve: First look at a graph of the data of speed vs. skid length and notice that it is not linear. It would be difficult to analyze. We added the point $(0,0)$ to the data table and graph because we are sure that if the speed were zero the skid length would also be zero.

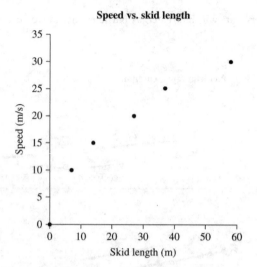

Speed vs. skid length

However, the theory has guided us to expect that a graph of speed **squared** vs. skid length would be linear and the slope would be $2a$. First we use a spreadsheet to square the speed and then graph the speed squared vs. skid length to see if it looks linear and that the intercept is close to zero. Only if it is linear is the deceleration constant, independent of speed.

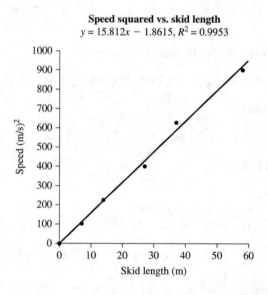

Speed squared vs. skid length
$y = 15.812x - 1.8615$, $R^2 = 0.9953$

(a) It looks linear and $R^2 = 0.995$ tells us the linear fit is very good. This means the deceleration (involved in the slope) is constant, independent of speed. We also see that the intercept is a very small negative number which is close to zero, so we have confidence in our model. The fit is not perfect and the intercept is not exactly zero probably because of uncertainties in measuring the speed.

(b) We now conclude that the slope of the best fit line $m = 15.812$ is $2a$ in the proper units, so the deceleration is $a = \frac{1}{2} \times 15.812 \text{ m/s}^2 = 7.9 \text{ m/s}^2$.

Assess: The value of 7.9 m/s^2 seems reasonable for hard braking. It is customary to put the independent variable on the horizontal axis and the dependent variable along the vertical axis. Had we done so here we would have graphed x vs. v^2 and the slope would have been $1/2a$. Our answer to the question would be the same.

VECTORS AND COORDINATE SYSTEMS

Exercises and Problems

Section 3.1 Vectors

Section 3.2 Properties of Vectors

3.1. Visualize:

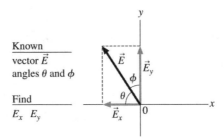

Solve: **(a)** To find $\vec{A}+\vec{B}$, we place the tail of vector $\vec{B}$ on the tip of vector $\vec{A}$ and draw an arrow from the tail of vector $\vec{A}$ to the tip of vector $\vec{B}$.

(b) Since $\vec{A}-\vec{B}=\vec{A}+(-\vec{B})$, we place the tail of the vector $-\vec{B}$ on the tip of vector $\vec{A}$ and then draw an arrow from the tail of vector $\vec{A}$ to the tip of vector $-\vec{B}$.

Section 3.3 Coordinate Systems and Vector Components

3.3. Visualize:

Solve: Vector $\vec{E}$ points to the left and up, so the components E_x and E_y are negative and positive, respectively, according to the Tactics Box 3.1.

(a) $E_x = -E\cos\theta$ and $E_y = E\sin\theta$.

(b) $E_x = -E\sin\varphi$ and $E_y = E\cos\varphi$.

Assess: Note that the role of sine and cosine are reversed because we are using a different angle. θ and ϕ are complementary angles.

3.7. Visualize:

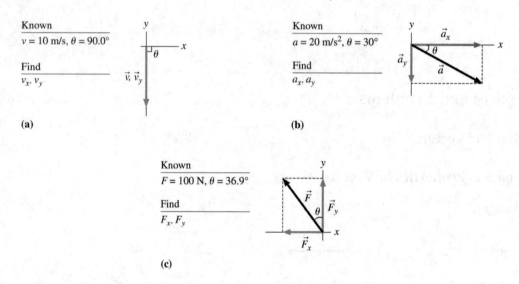

(a)   (b)

(c)

We will follow the rules given in Tactics Box 3.1.

Solve:

(a) $v_x = (10 \text{ m/s})\cos(90.0°) = 0 \text{ m/s}$ $v_y = -(10 \text{ cm/s})\sin(90.0°) = -10 \text{ m/s}$

(b) $a_x = (20 \text{ m/s}^2)\cos30° = 17 \text{ m/s}^2$ $a_y = -(20 \text{ m/s}^2)\sin30° = -10 \text{ m/s}^2$

(c) $F_x = -(100 \text{ N})\sin(36.9°) = -60 \text{ N}$ $F_y = (100 \text{ N})\cos(36.9°) = 80 \text{ N}$

Assess: The components have the same units as the vectors. Note the minus signs we have manually inserted according to Tactics Box 3.1.

Section 3.4 Vector Algebra

3.11. Visualize:

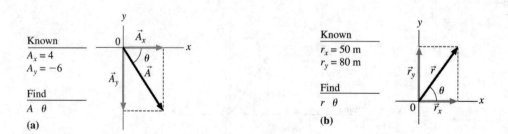

(a)   (b)

Solve: (a) Using the formulas for the magnitude and direction of a vector, we have:

$$A = \sqrt{(4)^2 + (-6)^2} = 7.2, \quad \theta = \tan^{-1}\left(\frac{r_y}{r_x}\right) = \tan^{-1}\left(\frac{6}{4}\right) = 56° \text{ below the } +x\text{-axis}$$

(b) $r = \sqrt{(50 \text{ m})^2 + (80 \text{ m})^2} = 94 \text{ m} \qquad \theta = \tan^{-1}\left(\frac{80 \text{ m}}{50 \text{ m}}\right) = 58° \text{ above the } +x\text{-axis}$

(c) $v = \sqrt{(-20 \text{ m/s})^2 + (40 \text{ m/s})^2} = 45 \text{ m/s} \qquad \theta = \tan^{-1}\left(\frac{40}{20}\right) = 63° \text{ above the } -x\text{-axis}$

(d) $a = \sqrt{(2.0 \text{ m/s}^2)^2 + (-6.0 \text{ m/s}^2)^2} = 6.3 \text{ m/s}^2 \qquad \theta = \tan^{-1}\left(\frac{2.0}{6.0}\right) = 18° \text{ to the right of the } -y\text{-axis}$

3.15. Visualize:

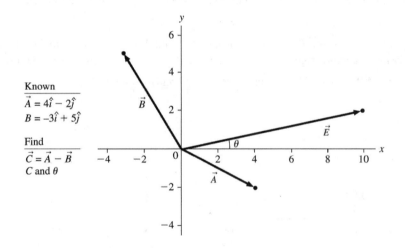

Solve: (a) We have $\vec{A} = 4\hat{i} - 2\hat{j}$ and $\vec{B} = -3\hat{i} + 5\hat{j}$, This means $4\vec{A} = 16\hat{i} - 8\hat{j}$ and $2\vec{B} = -6\hat{i} + 10\hat{j}$ Thus, $\vec{E} = 4\vec{A} + 2\vec{B} = [16 + (-6)]\hat{i} + [(-8) + 10]\hat{j} = 10\hat{i} + 2\hat{j}$.

(b) Vectors $\vec{A}$, $\vec{B}$, and $\vec{E}$ are shown in the figure above.

(c) From the $\vec{E}$ vector, $E_x = 10$ and $E_y = 2$. Therefore, the magnitude and direction of $\vec{E}$ are

$$E = \sqrt{(10)^2 + (2)^2} = \sqrt{104} = 10, \quad \theta = \tan^{-1}(E_y/E_x) = \tan^{-1}(2/10) = 11°$$

So $\vec{E}$ is 10, 11° above the +x-axis.

3.17. Visualize:

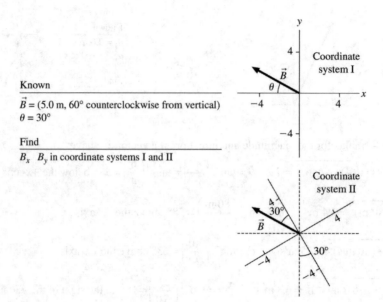

Known

$\vec{B}$ = (5.0 m, 60° counterclockwise from vertical)
$\theta = 30°$

Find

B_x B_y in coordinate systems I and II

Solve: In coordinate system I, the vector $\vec{B}$ makes an angle of 60° counterclockwise from vertical, so it has an angle of $\theta = 30°$ with the negative x-axis. Since $\vec{B}$ points to the left and up, it has a negative x-component and a positive y-component. Thus, $B_x = -(5.0 \text{ m})\cos(30°) = -4.3$ m and $B_y = +(5.0 \text{ m})\sin(30°) = 2.5$ m. Thus,

$$\vec{B} = -(4.3 \text{ m})\hat{i} + (2.5 \text{ m})\hat{j}.$$

In coordinate system II, the vector $\vec{B}$ makes an angle of 30° with the +y-axis and is to the left and up. This means we have to manually insert a minus sign for the x-component. Thus, $B_x = -B\sin(30°) = -(5.0 \text{ m})\sin(30°) = -2.5$ m, and $B_y = +B\cos(30°) = (5.0 \text{ m})\cos(30°) = 4.3$ m. Thus $\vec{B} = -(2.5 \text{ m})\hat{i} + (4.3 \text{ m})\hat{j}$.

3.19. Visualize: (a)

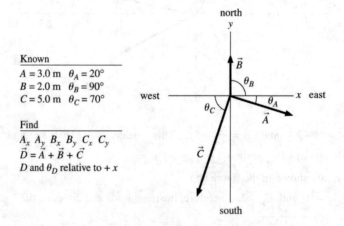

Known

$A = 3.0$ m $\theta_A = 20°$
$B = 2.0$ m $\theta_B = 90°$
$C = 5.0$ m $\theta_C = 70°$

Find

A_x A_y B_x B_y C_x C_y
$\vec{D} = \vec{A} + \vec{B} + \vec{C}$
D and θ_D relative to $+x$

Solve: (b) The components of the vectors $\vec{A}$, $\vec{B}$, and $\vec{C}$ are

$A_x = (3.0 \text{ m})\cos(20°) = 2.8$ m and $A_y = -(3.0 \text{ m})\sin(20°) = -1.0$ m; $B_x = 0$ m and $B_y = 2.0$ m;

$C_x = -(5.0 \text{ m})\cos(70°) = -1.7$ m and $C_y = -(5.0 \text{ m})\sin(70°) = -4.7$ m. This means the vectors can be written as

$$\vec{A} = (2.8\hat{i} + 1.0\hat{j}) \text{ m}, \quad \vec{B} = (2.0\hat{j}) \text{ m}, \quad \vec{C} = (-1.7\hat{i} - 4.7\hat{j}) \text{ m}$$

(c) We have $\vec{D} = \vec{A} + \vec{B} + \vec{C} = (1.1 \text{ m})\hat{i} - (3.7 \text{ m})\hat{j}$. This means

$$D = \sqrt{(1.1 \text{ m})^2 + (3.7 \text{ m})^2} = 3.9 \text{ m} \qquad \theta = \tan^{-1}(3.9/1.09) = 74°$$

The direction of $\vec{D}$ is south of east, 74° below the +x-axis.

3.23. Visualize: Refer to Figure P3.23 in your textbook.
Solve: (a) We are given that $\vec{A} + \vec{B} + \vec{C} = 1\hat{j}$ with $\vec{A} = 4\hat{i}$, and $\vec{C} = -2\hat{j}$. This means $\vec{A} + \vec{C} = 4\hat{i} - 2\hat{j}$. Thus, $\vec{B} = (\vec{A} + \vec{B} + \vec{C}) - (\vec{A} + \vec{C}) = (1\hat{j}) - (4\hat{i} - 2\hat{j}) = -4\hat{i} + 3\hat{j}$.

(b) We have $\vec{B} = B_x\hat{i} + B_y\hat{j}$ with $B_x = -4$ and $B_y = 3$. Hence, $B = \sqrt{(-4)^2 + (3)^2} = 5.0$

$$\theta = \tan^{-1}\frac{|B_y|}{|B_x|} = \tan^{-1}\left(\frac{3}{4}\right) = 37°$$

Since $\vec{B}$ has a negative x-component and a positive y-component, the vector $\vec{B}$ is in the second quadrant and the angle θ made by $\vec{B}$ is measured above the –x-axis.
Assess: Since $|B_y| < |B_x|$, $\theta < 45°$ as obtained above.

3.27. Visualize:

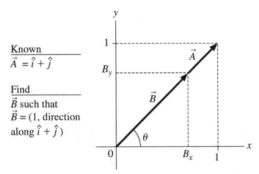

The magnitude of the unknown vector is 1 and its direction is along $\hat{i} + \hat{j}$. Let $\vec{A} = \hat{i} + \hat{j}$ as shown in the diagram. That is, $\vec{A} = 1\hat{i} + 1\hat{j}$ and the x- and y-components of $\vec{A}$ are both unity. Since $\theta = \tan^{-1}(A_y/A_x) = 45°$, the unknown vector must make an angle of 45° with the +x-axis and have unit magnitude.
Solve: Let the unknown vector be $\vec{B} = B_x\hat{i} + B_y\hat{j}$ where

$$B_x = B\cos(45°) = \frac{1}{\sqrt{2}}B \quad \text{and} \quad B_y = B\sin(45°) = \frac{1}{\sqrt{2}}B$$

We want the magnitude of $\vec{B}$ to be 1, so we have

$$B = \sqrt{B_x^2 + B_y^2} = 1 \Rightarrow \sqrt{\left(\frac{1}{\sqrt{2}}B\right)^2 + \left(\frac{1}{\sqrt{2}}B\right)^2} = 1 \Rightarrow \sqrt{B^2} = 1 \Rightarrow B = 1$$

Thus,

$$B_x = B_y = \frac{1}{\sqrt{2}}$$

Finally,

$$\vec{B} = B_x\hat{i} + B_y\hat{j} = \frac{1}{\sqrt{2}}\hat{i} + \frac{1}{\sqrt{2}}\hat{j}$$

3.29. Visualize: The coordinate system (x, y, z) is shown here; +x denotes east, +y denotes north, and +z denotes upward vertical. The vectors $\vec{S}_{\text{morning}}$ (shortened to $\vec{S}_m$), $\vec{S}_{\text{afternoon}}$ (shortened to $\vec{S}_a$), and the total displacement vector $\vec{S}_{\text{total}} = \vec{S}_a + \vec{S}_m$ are also shown.

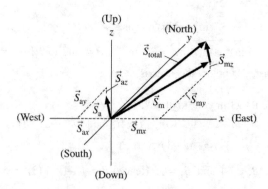

Solve: $\vec{S}_m = (2000\hat{i} + 3000\hat{j} + 200\hat{k})$ m, and $\vec{S}_a = (-1500\hat{i} + 2000\hat{j} - 300\hat{k})$ m. The total displacement is the sum of the individual displacements.

(a) The sum of the z-components of the afternoon and morning displacements is $S_{az} + S_{mz} = -300$ m $+ 200$ m $= -100$ m; that is, 100 m lower.

(b) $\vec{S}_{total} = \vec{S}_a + \vec{S}_m = (500\hat{i} + 5000\hat{j} - 100\hat{k})$ m; that is, (500 m east) + (5000 m north) − (100 m vertical). The magnitude of your total displacement is

$$S_{total} = \sqrt{(500)^2 + (5000)^2 + (-100)^2} \text{ m} = 5.0 \text{ km}$$

3.31. Visualize: (a)

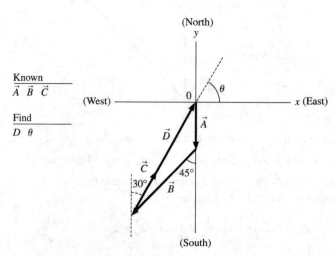

Note that $+x$ is along the east and $+y$ is along the north.

Solve: (b) We are given $\vec{A} = -(200 \text{ m})\hat{j}$, and can use trigonometry to obtain

$\vec{B} = -(400 \text{ m})\cos(45°) - (400 \text{ m})\sin(45°) = -(283 \text{ m})\hat{i} - (283 \text{ m})\hat{j}$ and

$\vec{C} = (200 \text{ m})\sin(30°) + (200 \text{ m})\cos(30°) = (100 \text{ m})\hat{i} + (173 \text{ m})\hat{j}$. We want $\vec{A} + \vec{B} + \vec{C} + \vec{D} = 0$, so

$\vec{D} = -\vec{A} - \vec{B} - \vec{C}$

$\quad = (200 \text{ m})\hat{j} - [-(283 \text{ m})\hat{i} - (283 \text{ m})\hat{j}] - [(100 \text{ m})\hat{i} + (173 \text{ m})\hat{j}] = (183 \text{ m})\hat{i} + (310 \text{ m})\hat{j}$

The magnitude and direction of $\vec{D}$ are

$$D = \sqrt{(183 \text{ m})^2 + (310 \text{ m})^2} = 360 \text{ m} \quad \text{and} \quad \theta = \tan^{-1}\left(\frac{D_y}{D_x}\right) = \tan^{-1}\left(\frac{310 \text{ m}}{183 \text{ m}}\right) = 59°$$

This means $\vec{D} = (360 \text{ m } 59° \text{ north of east})$.

(c) The measured length of the vector $\vec{D}$ on the graph (with a ruler) is approximately 1.75 times the measured length of vector $\vec{A}$. Since $A = 200$ m, this gives $D = 1.75 \times 200$ m $= 350$ m. Similarly, the angle θ measured with the protractor is close to $60°$. These answers are in close agreement to part (b).

3.35. Visualize:

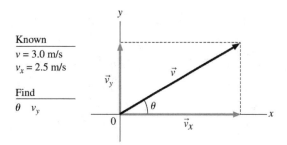

Known
v = 3.0 m/s
v_x = 2.5 m/s

Find
θ v_y

Solve: (a) Since $v_x = v\cos\theta$, we have 2.5 m/s $= (3.0 \text{ m/s})\cos\theta \Rightarrow \theta = \cos^{-1}\left(\dfrac{2.5 \text{ m/s}}{3.0 \text{ m/s}}\right) = 34°$.

(b) The vertical component is $v_y = v\sin\theta = (3.0 \text{ m/s})\sin(34°) = 1.7$ m/s.

3.37. Visualize:

(a)   **(b)**

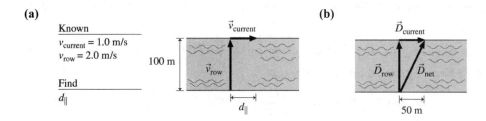

Known
$v_{current}$ = 1.0 m/s
v_{row} = 2.0 m/s

Find
$d_{\parallel}$

Solve: (a) The river is 100 m wide. If Mary rows due north at a constant speed of $v_{row} = 2.0$ m/s, it will take her (100 m)/(2.0 m/s) $= 50$ s to row across. But while she's doing so, the current sweeps her boat sideways at a speed $v_{current} = 1.0$ m/s. In the 50 s it takes her to cross the river, the current sweeps here a distance $d_{\parallel} = (v_{current} \times 50 \text{ s}) = 1.0 \text{ m/s} \times 50 \text{ s} = 50$ m, so she lands 50 m east of the point that was directly across the river from her when she started.

(b) Mary's net displacement $\vec{D}_{net}$, her displacement $\vec{D}_{current}$ due to the river's current, and her displacemnt $\vec{D}_{row}$ due to her rowing are shown in the figure.

3.39. Visualize: A 3% grade rises 3 m for every 100 m of horizontal distance. The angle of the ground is thus $\alpha = \tan^{-1}(3/100) = 1.72°$.

Establish a tilted coordinate system with one axis parallel to the ground and the other axis perpendicular to the ground.

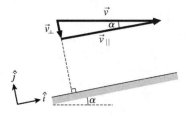

Solve: From the figure, the magnitude of the component vector of $\vec{v}$ perpendicular to the ground is $v_\perp = v\sin\alpha = 15.0$ m/s. But this is only the size. We also have to note that the *direction* of $\vec{v}_\perp$ is down, so the component is $\vec{v}_\perp = -(15 \text{ m/s})\hat{j}$.

3.43. Visualize:

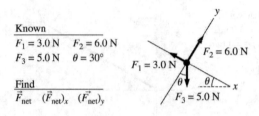

Known
$F_1 = 3.0$ N $F_2 = 6.0$ N
$F_3 = 5.0$ N $\theta = 30°$

Find
$\vec{F}_{net}$ $(\vec{F}_{net})_x$ $(\vec{F}_{net})_y$

Use a tilted coordinate system such that x-axis is down the slope.
Solve: Expressing all three forces in terms of unit vectors, we have $\vec{F}_1 = -(3.0 \text{ N})\hat{i}$, $\vec{F}_2 = +(6.0 \text{ N})\hat{j}$, and $\vec{F}_3 = (5.0 \text{ N})\sin\theta\,\hat{i} - (5.0 \text{ N})\cos\theta\,\hat{j}$.

(a) The component of $\vec{F}_{net}$ parallel to the floor is $(F_{net})_x = -(3.0 \text{ N}) + 0 \text{ N} + (5.0 \text{ N})\sin(30°) = -0.50$ N, or 0.50 N up the slope.

(b) The component of $\vec{F}_{net}$ perpendicular to the floor is $(F_{net})_y = 0 \text{ N} + (6.0 \text{ N}) - (5.0 \text{ N})\cos(30°) = 1.67$ N, or 1.7 N to two significant figures.

(c) The magnitude of $\vec{F}_{net}$ is $F_{net} = \sqrt{(F_{net})_x + (F_{net})_y} = \sqrt{(-0.50 \text{ N})^2 + (1.67 \text{ N})^2} = 1.74$ N, or 1.7 N to two significant figures. The angle between $\vec{F}_{net}$ and the negative x-axis is

$$\phi = \tan^{-1}\frac{(F_{net})_y}{\left|(F_{net})_x\right|} = \tan^{-1}\left(\frac{1.67 \text{ N}}{0.50 \text{ N}}\right) = 73°$$

$\vec{F}_{net}$ is $73°$ clockwise from the $-x$-axis.

KINEMATICS IN TWO DIMENSIONS

Exercises and Problems

Section 4.1 Acceleration

4.3. Solve: To make the particle slow down the acceleration needs to have a component that is opposite the direction of the velocity. To make the particle curve upward the acceleration must have a component upward. So the answer is H.

4.5. Solve: To make the particle speed up the acceleration needs to have a component that is in the direction of the velocity. To make the particle curve downward the acceleration must have a component downward. So the answer is E.

Section 4.2 Two-Dimensional Kinematics

4.9. Model: The puck is a particle and follows the constant-acceleration kinematic equations of motion.
Solve: (a) At $t = 2$ s, the graphs give $v_x = 16$ cm/s and $v_y = 30$ cm/s. The angle made by the vector $\vec{v}$ with the x-axis can thus be found as

$$\theta = \tan^{-1}\left(\frac{v_y}{v_x}\right) = \tan^{-1}\left(\frac{30 \text{ cm/s}}{16 \text{ cm/s}}\right) = 62° \text{ above the } x\text{-axis}$$

(b) After $t = 5$ s, the puck has traveled a distance given by:

$$x_1 = x_0 + \int_0^{5s} v_x \, dt = 0 \text{ m} + \text{area under } v_x\text{-}t \text{ curve} = \tfrac{1}{2}(40 \text{ cm/s})(5 \text{ s}) = 100 \text{ cm}$$

$$y_1 = y_0 + \int_0^{5s} v_y \, dt = 0 \text{ m} + \text{area under } v_y\text{-}t \text{ curve} = (30 \text{ cm/s})(5 \text{ s}) = 150 \text{ cm}$$

$$\Rightarrow r_1 = \sqrt{x_1^2 + y_1^2} = \sqrt{(100 \text{ cm})^2 + (150 \text{ cm})^2} = 180 \text{ cm}$$

Section 4.3 Projectile Motion

4.11. Model: Assume the particle model for the ball, and apply the constant-acceleration kinematic equations of motion in a plane.

Visualize:

Pictorial representation

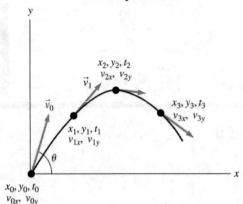

Known
$t_1 = 1$ s
$x_0 = y_0 = t_0 = 0$
$\vec{v}_1 = (2.0\hat{i} + 2.0\hat{j})$ m/s
$v_{1x} = 2.0$ m/s
$y_{1y} = 2.0$ m/s

Find
v_{0x} v_{0y}
v_{2x} v_{2y}
v_{3x} v_{3y}
$a = g$
θ

Solve: (a) We know the velocity $\vec{v}_1 = (2.0\hat{i} + 2.0\hat{j})$ m/s at $t = 1$ s. The ball is at its highest point at $t = 2$ s, so $v_y = 0$ m/s. The horizontal velocity is constant in projectile motion, so $v_x = 2.0$ m/s at all times. Thus $\vec{v}_2 = 2.0\hat{i}$ m/s at $t = 2$ s. We can see that the y-component of velocity *changed* by $\Delta v_y = -2.0$ m/s between $t = 1$ s and $t = 2$ s. Because a_y is constant, v_y changes by -2.0 m/s in *any* 1-s interval. At $t = 3$ s, v_y is 2.0 m/s less than its value of 0 at $t = 2$ s. At $t = 0$ s, v_y must have been 2.0 m/s more than its value of 2.0 m/s at $t = 1$ s. Consequently, at $t = 0$ s,

$$\vec{v}_0 = (2.0\hat{i} + 4.0\hat{j}) \text{ m/s}$$

At $t = 1$ s,

$$\vec{v}(1) = (2.0\hat{i} + 2.0\hat{j}) \text{ m/s}$$

At $t = 2$ s,

$$\vec{v}(2) = (2.0\hat{i} + 0.0\hat{j}) \text{ m/s}$$

At $t = 3$ s,

$$\vec{v}(3) = (2.0\hat{i} - 2.0\hat{j}) \text{ m/s}$$

(b) Because v_y is changing at the rate -2.0 m/s per s, the y-component of acceleration is $a_y = -2.0$ m/s². But $a_y = -g$ for projectile motion, so the value of g on Exidor is $g = 2.0$ m/s².

(c) From part (a) the components of $\vec{v}_0$ are $v_{0x} = 2.0$ m/s and $v_{0y} = 4.0$ m/s. This means

$$\theta = \tan^{-1}\left(\frac{v_{0y}}{v_{0x}}\right) = \tan^{-1}\left(\frac{4.0 \text{ m/s}}{2.0 \text{ m/s}}\right) = 63° \text{ above } +x$$

Assess: The y-component of the velocity vector decreases from 2.0 m/s at $t = 1$ s to 0 m/s at $t = 2$ s. This gives an acceleration of -2 m/s². All the other values obtained above are also reasonable.

4.13. Model: The bullet is treated as a particle and the effect of air resistance on the motion of the bullet is neglected.
Visualize:

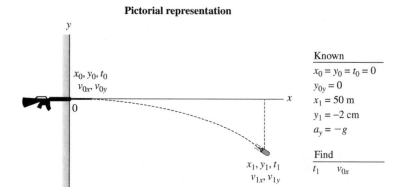

Pictorial representation

Known
$x_0 = y_0 = t_0 = 0$
$v_{0y} = 0$
$x_1 = 50$ m
$y_1 = -2$ cm
$a_y = -g$

Find
t_1 v_{0x}

Solve: (a) Using $y_1 = y_0 + v_{0y}(t_1 - t_0) + \frac{1}{2}a_y(t_1 - t_0)^2$, we obtain

$$(-2.0 \times 10^{-2} \text{ m}) = 0 \text{ m} + 0 \text{ m} + \frac{1}{2}(-9.8 \text{ m/s}^2)(t_1 - 0 \text{ s})^2 \Rightarrow t_1 = 0.0639 \text{ s} \approx 0.064 \text{ s}$$

(b) Using $x_1 = x_0 + v_{0x}(t_1 - t_0) + \frac{1}{2}a_x(t_1 - t_0)^2$,

$$(50 \text{ m}) = 0 \text{ m} + v_{0x}(0.0639 \text{ s} - 0 \text{ s}) + 0 \text{ m} \Rightarrow v_{0x} = 782 \text{ m/s} \approx 780 \text{ m/s}$$

Assess: The bullet falls 2 cm during a horizontal displacement of 50 m. This implies a large initial velocity, and a value of 782 m/s is understandable.

Section 4.4 Relative Motion

4.15. Model: Assume motion along the x-direction (downstream to the right). Call the speed of the boat with respect to the water $(v_x)_{BW}$, the speed of the water with respect to the Earth $(v_x)_{WE}$, and the speed of the boat with respect to the Earth $(v_x)_{BE}$.
Solve: We seek $(v_x)_{WE}$.

Downstream: $(v_x)_{BE} = (v_x)_{BW} + (v_x)_{WE} = \dfrac{30 \text{ km}}{3.0 \text{ h}} = 10 \text{ km/h}$

Upstream: $(v_x)_{BE} = -(v_x)_{BW} + (v_x)_{WE} = \dfrac{30 \text{ km}}{5.0 \text{ h}} - -6.0 \text{ km/h}$

Add the two equations to get $2(v_x)_{WE} = 4.0$ km/h, so the river flows at 2.0 m/s.

Assess: This means that the boat goes at 8.0 m/s relative to the water. Both these numbers sound reasonable.

4.21. Solve: The angular position graph is the area under the angular velocity graph. At $t = 4$ s the area is 80 rad. Between 4 s and 6 s the angular velocity is zero so the angular position doesn't change. Between 6 s and 8 s the area is 20 rad, but it is below the axis, so we subtract it. The area under the ω versus t graph during the total time interval of 8 s is 80 rad -20 rad $= 60$ rad. This is where we end up on the θ axis at 8.s

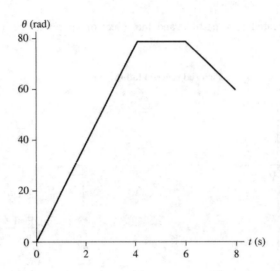

Section 4.5 Uniform Circular Motion

4.23. Model: The airplane is to be treated as a particle.
Visualize:

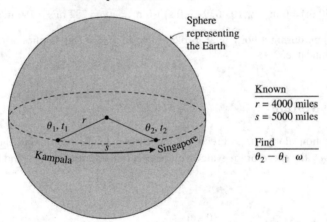

Pictorial representation

Sphere
representing
the Earth

Known
$r = 4000$ miles
$s = 5000$ miles

Find
$\theta_2 - \theta_1$ ω

Solve: The angle you turn through is

$$\theta_2 - \theta_1 = \frac{s}{r} = \frac{5000 \text{ miles}}{4000 \text{ miles}} = 1.2500 \text{ rad } = 1.2500 \text{ rad} \times \frac{180°}{\pi \text{ rad}} = 71.62°$$

The plane's angular velocity is

$$\omega = \frac{\theta_2 - \theta_1}{t_2 - t_1} = \frac{71.62°}{9 \text{ h}} = 8.0°/\text{h}$$

Section 4.6 Velocity and Acceleration in Uniform Circular Motion

4.25. Solve: The plane must fly as fast as the earth's surface moves, but in the opposite direction. That is, the plane must fly from east to west. The speed is

$$v = \omega r = \left(\frac{2\pi \text{ rad}}{24 \text{ h}}\right)(6.4\times10^3 \text{ km}) = 1680 \frac{\text{km}}{\text{h}} = 1680 \ \frac{\text{km}}{\text{h}} \times \frac{1 \text{ mile}}{1.609 \text{ km}} = 1040 \text{ mph}$$

4.29. Model: The crankshaft is a rotating rigid body.
Visualize: The angular acceleration is the slope of the angular velocity graph.

Solve:

(a) The slope of the graph at $t = 1$ s is $\dfrac{-200 \text{ rad/s}}{2 \text{ s}} = -100 \text{ rad/s}^2$.

(b) The slope of the graph at $t = 3$ s is 0 rad/s^2.

(c) The slope of the graph at $t = 1$ s is $\dfrac{150 \text{ rad/s}}{3 \text{ s}} = 50 \text{ rad/s}^2$.

4.31. Visualize: The angular position is the slope of the area under the ω vs. t graph.
Solve: The area under the graph is $20 \text{ rad} + 40 \text{ rad} = 60 \text{ rad}$. Convert to revolutions. $60 \text{ rad}(1 \text{ rev}/2\pi \text{ rad}) = 9.5 \text{ rev}$.

4.33. Model: The fan is in nonuniform circular motion.
Visualize:

Pictorial representation

Known
$\omega_i = 0$
$\omega_f = 1800$ rpm
$\Delta t = 4$ s

Find
α

Solve: Note $1800 \text{ rev/min}\left(\dfrac{\text{min}}{60 \text{ s}}\right) = 30 \text{ rev/s}$. Thus $\omega_f = \omega_i + \alpha \Delta t \Rightarrow 30 \text{ rev/s} = 0 \text{ rev/s} + \alpha(4.0 \text{ s}) \Rightarrow \alpha = 7.5 \text{ rev/s}^2$.

This can be expressed as $(7.5 \text{ rev/s})\left(\dfrac{2\pi \text{ rad}}{\text{rev}}\right) = 47 \text{ rad/s}^2$.

Assess: An increase in the angular velocity of a fan blade by 7.5 rev/s each second seems reasonable.

4.39. Model: Assume particle motion in a plane and constant-acceleration kinematics for the projectile.
Visualize:

Pictorial representation

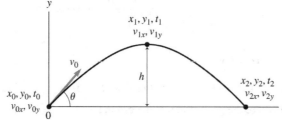

Known
v_0 $\quad\theta$ $\quad v_{1y}$
$x_0 = y_0 = t_0 = 0$

Find
$h \,(= y_1 - y_0)$ and
$x_2 - x_1$ for $\theta =$
$30°$, $45°$ and $60°$

Solve: (a) We know that $v_{0y} = v_0 \sin\theta$, $a_y = -g$, and $v_{1y} = 0$ m/s. Using $v_{1y}^2 = v_{0y}^2 + 2a_y(y_1 - y_0)$,

$$0 \text{ m}^2/\text{s}^2 = v_0^2 \sin^2\theta + 2(-g)h \Rightarrow h = \frac{v_0^2 \sin^2\theta}{2g}$$

(b) Using the equation for range and the above expression for $\theta = 30.0°$:

$$h = \frac{(33.6 \text{ m/s})^2 \sin^2 30.0°}{2(9.8 \text{ m/s}^2)} = 14.4 \text{ m}$$

$$(x_2 - x_0) = \frac{v_0^2 \sin 2\theta}{g} = \frac{(33.6 \text{ m/s})^2 \sin(2 \times 30.0°)}{(9.8 \text{ m/s}^2)} = 99.8 \text{ m}$$

For $\theta = 45.0°$:

$$h = \frac{(33.6 \text{ m/s})^2 \sin^2 45.0°}{2(9.8 \text{ m/s}^2)} = 28.8 \text{ m}$$

$$(x_2 - x_0) = \frac{(33.6 \text{ m/s})^2 \sin(2 \times 45.0°)}{(9.8 \text{ m/s}^2)} = 115.2 \text{ m}$$

For $\theta = 60.0°$:

$$h = \frac{(33.6 \text{ m/s})^2 \sin^2 60.0°}{2(9.8 \text{ m/s}^2)} = 43.2 \text{ m}$$

$$(x_2 - x_0) = \frac{(33.6 \text{ m/s})^2 \sin(2 \times 60.0°)}{2(9.8 \text{ m/s}^2)} = 99.8 \text{ m}$$

Assess: The projectile's range, being proportional to $\sin(2\theta)$, is maximum at a launch angle of 45°, but the maximum height reached is proportional to $\sin^2(\theta)$. These dependencies are seen in this problem.

4.41. Model: Assume the particle model for the projectile and motion in a plane.
Visualize:

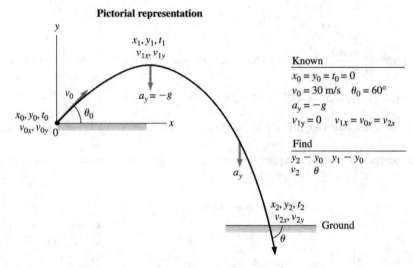

Pictorial representation

Known
$x_0 = y_0 = t_0 = 0$
$v_0 = 30 \text{ m/s} \quad \theta_0 = 60°$
$a_y = -g$
$v_{1y} = 0 \quad v_{1x} = v_{0x} = v_{2x}$

Find
$y_2 - y_0 \quad y_1 - y_0$
$v_2 \quad \theta$

Solve: (a) Using $y_2 = y_0 + v_{0y}(t_2 - t_0) + \frac{1}{2} a_y (t_2 - t_0)^2$,

$$y_2 = 0 \text{ m} + (30 \text{ m/s}) \sin 60°(7.5 \text{ s} - 0 \text{ s}) + \frac{1}{2}(-9.8 \text{ m/s}^2)(7.5 \text{ s} - 0 \text{ s})^2 = -80.8 \text{ m}$$

Thus the launch point is 81 m higher than where the projectile hits the ground.
(b) Using $v_{1y}^2 = v_{0y}^2 + 2a_y(y_1 - y_0)$,

$$0 \text{ m}^2/\text{s}^2 = (30 \sin 60° \text{ m/s})^2 + 2(-9.8 \text{ m/s}^2)(y_1 - 0 \text{ m}) \Rightarrow y_1 = 34.4 \text{ m, or } y_1 = 34 \text{ m}$$

Assess: The projectile hits the ground at an angle of 73°.

4.45. Model: The particle model for the ball and the constant-acceleration equations of motion in a plane are assumed.
Visualize:

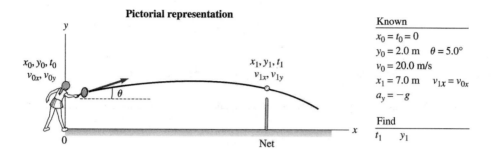

Pictorial representation

Known
$x_0 = t_0 = 0$
$y_0 = 2.0$ m $\theta = 5.0°$
$v_0 = 20.0$ m/s
$x_1 = 7.0$ m $v_{1x} = v_{0x}$
$a_y = -g$

Find
t_1 y_1

Solve: The initial velocity is
$$v_{0x} = v_0 \cos 5.0° = (20 \text{ m/s}) \cos 5.0° = 19.92 \text{ m/s}$$
$$v_{0y} = v_0 \sin 5.0° = (20 \text{ m/s}) \sin 5.0° = 1.743 \text{ m/s}$$
The time it takes for the ball to reach the net is
$$x_1 = x_0 + v_{0x}(t_1 - t_0) \Rightarrow 7.0 \text{ m} = 0 \text{ m} + (19.92 \text{ m/s})(t_1 - 0 \text{ s}) \Rightarrow t = 0.351 \text{ s}$$
The vertical position at $\vec{v} = \vec{v} + \vec{V}$ is
$$y_1 = y_0 + v_{0y}(t_1 - t_0) + \frac{1}{2} a_y (t_1 - t_0)^2$$
$$= (2.0 \text{ m}) + (1.743 \text{ m/s})(0.351 \text{ s} - 0 \text{ s}) + \frac{1}{2}(-9.8 \text{ m/s}^2)(0.351 \text{ s} - 0 \text{ s})^2 = 2.01 \text{ m}$$
Thus the ball clears the net by $1.01 \text{ m} \approx 1.0 \text{ m}$.
Assess: The vertical free fall of the ball, with zero initial velocity, in 0.351 s is 0.6 m. The ball will clear by approximately 0.4 m if the ball is thrown horizontally. The initial launch angle of 5° provides some initial vertical velocity and the ball clears by a larger distance. The above result is reasonable.

4.47. Model: We will use the particle model and the constant-acceleration kinematic equations in a plane.
Visualize:

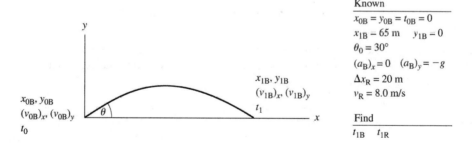

Pictorial representation

Known
$x_{0B} = y_{0B} = t_{0B} = 0$
$x_{1B} - 65$ m $y_{1B} - 0$
$\theta_0 = 30°$
$(a_B)_x = 0$ $(a_B)_y = -g$
$\Delta x_R = 20$ m
$v_R = 8.0$ m/s

Find
t_{1B} t_{1R}

Solve: The x-and y-equations of the ball are
$$x_{1B} = x_{0B} + (v_{0B})_x(t_{1B} - t_{0B}) + \frac{1}{2}(a_B)_x(t_{1B} - t_{0B})^2 \Rightarrow 65 \text{ m} = 0 \text{ m} + (v_{0B}\cos 30°)t_{1B} + 0 \text{ m}$$
$$y_{1B} = y_{0B} + (v_{0B})_y(t_{1B} - t_{0B}) + \frac{1}{2}(a_B)_y(t_{1B} - t_{0B})^2 \Rightarrow 0 \text{ m} = 0 \text{ m} + (v_{0B}\sin 30°)t_{1B} + \frac{1}{2}(-g)t_{1B}^2$$
From the y-equation,
$$v_{0B} = \frac{gt_{1B}}{(2\sin 30°)}$$

Substituting this into the x-equation yields

$$65 \text{ m} = \frac{g \cos 30° \, t_{1B}^2}{2 \sin 30°}$$

$$\Rightarrow t_{1B} = 2.77 \text{ s}$$

For the runner:

$$t_{1R} = \frac{20 \text{ m}}{8.0 \text{ m/s}} = 2.50 \text{ s}$$

Thus, the throw is too late by 0.27 s.

Assess: The times involved in running the bases are small, and a time of 2.5 s is reasonable.

4.49. Model: We will assume a particle model for the sand, and use the constant-acceleration kinematic equations.

Visualize:

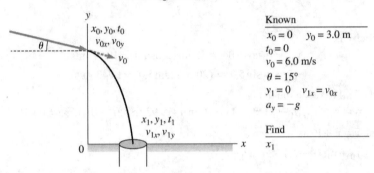

Pictorial representation

Known
$x_0 = 0$ $y_0 = 3.0$ m
$t_0 = 0$
$v_0 = 6.0$ m/s
$\theta = 15°$
$y_1 = 0$ $v_{1x} = v_{0x}$
$a_y = -g$

Find
x_1

Solve: Using the equation $x_1 = x_0 + v_{0x}(t_1 - t_0) + \frac{1}{2}a_x(t_1 - t_0)^2$,

$$x_1 = 0 \text{ m} + (v_0 \cos 15°)(t_1 - 0 \text{ s}) + 0 \text{ m} = (60 \text{ m/s})(\cos 15°)t_1$$

We can find t_1 from the y-equation, but note that $v_{0y} = -v_0 \sin 15°$ because the sand is launched at an angle below horizontal.

$$y_1 = y_0 + v_{0y}(t_1 - t_0) + \frac{1}{2}a_y(t_1 - t_0)^2 \Rightarrow 0 \text{ m} = 3.0 \text{ m} - (v_0 \sin 15°)t_1 - \frac{1}{2}gt_1^2$$

$$= 3.0 \text{ m} - (6.0 \text{ m/s})(\sin 15°)t_1 - \frac{1}{2}(9.8 \text{ m/s}^2)t_1^2$$

$$\Rightarrow 4.9t_1^2 + 1.55t_1 - 3.0 = 0 \Rightarrow t_1 = 0.6399 \text{ s and} -0.956 \text{ s (unphysical)}$$

Substituting this value of t_1 in the x-equation gives the distance

$$d = x_1 = (6.0 \text{ m/s})\cos 15°(0.6399 \text{ s}) = 3.71 \text{ m} \approx 3.7 \text{ m}$$

4.51. Model: Assuming constant acceleration allows us to use the kinematic equations.

Visualize: We apply the kinematic equations during the free-fall flight to find the velocity as the javelin left the hand. Then use $v_f^2 = v_i^2 + 2a_s \Delta s$ where $\Delta s = 0.70$m.

Solve: The range is $\Delta x = 62$m.

$$\Delta x = (v_0)_x \Delta t = v_0 \cos \theta \Delta t \Rightarrow \Delta t = \frac{\Delta x}{v_0 \cos \theta}$$

$$y_f = y_i + (v_0 \sin \theta)\Delta t + \frac{1}{2}a_y(\Delta t)^2$$

Insert our new expression for Δt.

$$\Delta y = (v_0 \sin \theta)\frac{\Delta x}{v_0 \cos \theta} + \frac{1}{2}(-g)\left(\frac{\Delta x}{v_0 \cos \theta}\right)^2$$

Solve for v_0.

$$\Delta y = (\tan\theta)\Delta x + \frac{1}{2}(-g)\left(\frac{\Delta x}{v_0\cos\theta}\right)^2$$

$$\frac{1}{2}(g)\left(\frac{\Delta x}{v_0\cos\theta}\right)^2 = (\tan\theta)\Delta x - \Delta y$$

$$v_0^2 = \frac{g}{2}\left(\frac{\Delta x}{\cos\theta}\right)^2\left(\frac{1}{(\tan\theta)\Delta x - \Delta y}\right)$$

$$v_0 = \sqrt{\frac{g}{2}\left(\frac{\Delta x}{\cos\theta}\right)^2\left(\frac{1}{(\tan\theta)\Delta x - \Delta y}\right)}$$

$$= \sqrt{\frac{9.8\ \text{m/s}^2}{2}\left(\frac{62\ \text{m}}{\cos 30°}\right)^2\left(\frac{1}{(\tan 30°)(62\ \text{m}) - (-2\ \text{m})}\right)} = 25.78\ \text{m/s}$$

This is the speed as the javelin leaves the hand. It now becomes v_f as we consider the time during the throw (as the hand accelerates it from rest).

$$a = \frac{v_f^2 - v_i^2}{2\Delta s} = \frac{(25.78\ \text{m/s})^2 - (0\ \text{m/s})^2}{2(0.70\ \text{m})} = 470\ \text{m/s}^2$$

Assess: This is a healthy acceleration, but what is required for a good throw.

4.55. Model: Use subscripts C, R, and G for car, rain, and ground respectively.
Solve: The Galilean transformation of velocity is $\vec{v}_{RG} = \vec{v}_{RC} + \vec{v}_{CG}$. While driving north, $\vec{v}_{CG} = (25\ \text{m/s})\hat{i}$ and $v_{RG} = -v_R\cos\theta\,\hat{j} - v_R\sin\theta\,\hat{i}$. Thus,

$$\vec{v}_{RC} = \vec{v}_{RG} - \vec{v}_{CG} = (-v_R\sin\theta - 25\ \text{m/s})\hat{i} - v_R\cos\theta\,\hat{j}$$

Since the observer in the car finds the raindrops making an angle of 38° with the vertical, we have

$$\frac{v_R\sin\theta + 25\ \text{m/s}}{v_R\cos\theta} = \tan 38°$$

While driving south, $\vec{v}_{CG} = -(25\ \text{m/s})\hat{i}$, and $\vec{v} = -v_R\cos\theta\,\hat{j} - v_R\sin\theta\,\hat{i}$. Thus,

$$\vec{v}_{RG} = (-v_R\sin\theta + 25\ \text{m/s})\hat{i} - v_R\cos\theta\,\hat{j}$$

Since the observer in the car finds the raindrops falling vertically straight, we have

$$\frac{-v_R\sin\theta + 25\ \text{m/s}}{v_R\cos\theta} = \tan 0° = 0 \Rightarrow v_R\sin\theta = 25\ \text{m/s}$$

Substituting this value of $v_R\sin\theta$ into the expression obtained for driving north yields:

$$\frac{25\ \text{m/s} + 25\ \text{m/s}}{v_R\cos\theta} = \tan 38° \Rightarrow v_R\cos\theta = \frac{50\ \text{m/s}}{\tan 38°} = 64.0\ \text{m/s}$$

Therefore, we have for the velocity of the raindrops:

$$(v_R\sin\theta)^2 + (v_R\cos\theta)^2 = (25\ \text{m/s})^2 + (64.0\ \text{m/s})^2 \Rightarrow v_R^2 = 4721(\text{m/s})^2 \Rightarrow v_R = 68.7\ \text{m/s}$$

$$\tan\theta = \frac{v_R\sin\theta}{v_R\cos\theta} = \frac{25\ \text{m/s}}{64\ \text{m/s}} \Rightarrow \theta = 21.3°$$

The raindrops fall at 69 m/s while making an angle of 21° with the vertical.

4.59. Model: We will use the particle model for the astronaut undergoing nonuniform circular motion.

Solve: **(a)** The initial conditions are $\omega_0 = 0$ rad/s, $\theta_0 = 0$ rad, $t_0 = 0$ s, and $r = 6.0$ m. After 30 s,

$$\omega_1 = \frac{1 \text{ rev}}{1.3 \text{ s}} = \frac{1}{1.3} \times \frac{\text{rev}}{\text{s}} \times \frac{2\pi \text{ rad}}{\text{rev}} = 4.83 \text{ rad/s}$$

Using these values at $t_1 = 30$ s,

$$\omega_1 = \omega_0 + (a_t/r)(t_1 - t_0) = 0 + (a_t/r)t_1$$

$$\Rightarrow a_t = (6.0 \text{ m})(4.83 \text{ rad/s})\left(\frac{1}{30 \text{ s}}\right) = 0.97 \text{ m/s}^2$$

(b) The radial acceleration is

$$a_r = r\omega_1^2 = (6.0 \text{ m})(4.83 \text{ rad/s})^2 \frac{g}{(9.8 \text{ m/s}^2)} = 14.3g \approx 14g$$

Assess: The above acceleration is typical of what astronauts experience during liftoff.

4.61. Model: The earth is a rigid, rotating, and spherical body.

Visualize:

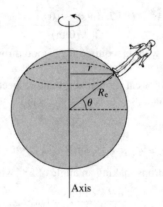

Solve: At a latitude of θ degrees, the radius is $r = R_e \cos\theta$ with $R_e = 6400$ km $= 6.400 \times 10^6$ m.

(a) In Miami $\theta = 26°$, and we have $r = (6.400 \times 10^6 \text{ m})(\cos 26°) = 5.752 \times 10^6$ m. The angular velocity of the earth is

$$\omega = \frac{2\pi}{T} = \frac{2\pi}{24 \times 3600 \text{ s}} = 7.272 \times 10^{-5} \text{ rad/s}$$

Thus, $v_{\text{student}} = r\omega = (5.752 \times 10^6 \text{ m})(7.272 \times 10^{-5} \text{ rad/s}) = 418 \text{ m/s} \approx 420$ m/s.

(b) In Fairbanks $\theta = 65°$, so $r = (6.400 \times 10^6 \text{ m})\cos 65° = 2.705 \times 10^6$ m and $v_{\text{student}} = r\omega = (2.705 \times 10^6 \text{ m})$

$(7.272 \times 10^{-5} \text{ rad/s}) = 197 \text{ m/s} \approx 200$ m/s.

4.65. Model: The drill is a rigid rotating body.

Visualize:

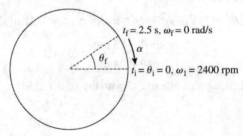

The figure shows the drill's motion from the top.

Solve: (a) The kinematic equation $\omega_f = \omega_i + \alpha(t_f - t_i)$ becomes, after using

$\omega_i = 2400$ rpm $= (2400)(2\pi)/60 = 251.3$ rad/s, $t_f - t_i = 2.5$ s $- 0$ s $= 2.5$ s, and $w_f = 0$ rad/s,

$$0 \text{ rad} = 251.3 \text{ rad/s} + \alpha(2.5 \text{ s}) \Rightarrow \alpha = -100 \text{ rad/s}^2$$

(b) Applying the kinematic equation for angular position yields:

$$\theta_f = \theta_i + \omega_i(t_f - t_i) + \frac{1}{2}\alpha(t_f - t_i)^2$$

$$= 0 \text{ rad} + (251.3 \text{ rad/s})(2.5 \text{ s} - 0 \text{ s}) + \frac{1}{2}(-100 \text{ rad/s}^2)(2.5 \text{ s} - 0 \text{ s})^2$$

$$= 3.2 \times 10^2 \text{ rad} = 50 \text{ rev}$$

4.67. Model: Model the tire as a rotating rigid body. Assume the angular acceleration is constant. The radius of the tire is 32 cm.
Visualize: $\omega_i = 3.5$ rev/s $= 22$ rad/s; $\omega_f = 6.0$ rev/s $= 37.7$ rad/s.

Solve: $\Delta\theta = \dfrac{\Delta x}{r}$.

$$\alpha = \frac{\omega_f^2 - \omega_i^2}{2\Delta\theta} = \frac{\omega_f^2 - \omega_i^2}{2\dfrac{\Delta x}{r}} = \frac{(37.7 \text{ rad/s})^2 - (22 \text{ rad/s})^2}{2\dfrac{200 \text{ m}}{0.32 \text{ m}}} = 0.75 \text{ rad/s}^2$$

Assess: The units all check out.

4.71. Model: Model the shaft as a rotating rigid body. Assume the angular acceleration is constant.
Visualize: The data gives angular velocity as a function of time. If the angular acceleration is constant it will be the slope of the best-fit straight line through the data. First convert the data from rpm to rad/s.

t (s)	ω (rpm)	ω (rad/s)
0	3010	315.21
1	2810	294.26
2	2450	256.56
3	2250	235.62
4	1940	203.16
5	1810	189.54
6	1510	158.13

Solve: The slope of the linear regression line is the angular acceleration.

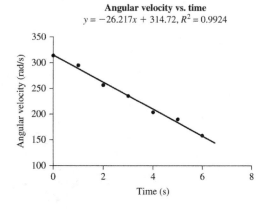

Angular velocity vs. time
$y = -26.217x + 314.72, R^2 = 0.9924$

The spreadsheet says the slope is $\alpha = -26.22$ rad/s^2 ≈ -26 rad/s^2. The magnitude is 26 rad/s^2.

Assess: It's hard to have an intuitive feel for reasonable values of α, but our answer doesn't seem ridiculous.

4.73. Model: The string is wrapped around the spool in such a way that it does not pile up on itself, and unwinds without slipping.

Visualize:

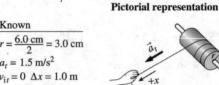

Pictorial representation

Known

$r = \dfrac{6.0 \text{ cm}}{2} = 3.0$ cm

$a_t = 1.5$ m/s^2

$v_{it} = 0 \quad \Delta x = 1.0$ m

Find

ω_f

Solve: Since the string unwinds without slipping, the angular distance the spool turns as the string is pulled 1.0 m is

$$\Delta \theta = \frac{\Delta x}{r} = \frac{1.0 \text{ m}}{3.0 \times 10^{-2} \text{ m}} = 33 \text{ radians.}$$

The angular acceleration of the spool due to the pull on the string is

$$\alpha = \frac{a_t}{r} = \frac{1.5 \text{ m/s}^2}{3.0 \times 10^{-2} \text{ m}} = 50 \text{ rad/s}^2$$

The angular velocity of the spool after pulling the string is found with kinematics.

$$\omega_f^2 = \omega_i^2 + 2\alpha\Delta\theta \Rightarrow \omega_f^2 = 0 \text{ rad}^2/\text{s}^2 + 2(50 \text{ rad/s}^2)(33 \text{ rad})$$

$$\Rightarrow \omega_f^2 = 57 \text{ rad/s}$$

Converting to revolutions per minute,

$$(57 \text{ rad/s})\left(\frac{\text{rev}}{2\pi \text{ rad}}\right)\left(\frac{60 \text{ s}}{\text{min}}\right) = 5.5 \times 10^2 \text{ rpm}$$

Assess: The angular speed of 57 rad/s $\approx$ 9 rev/s is reasonable for a medium-sized spool.

FORCE AND MOTION

Exercises and Problems

Section 5.3 Identifying Forces

5.1. Visualize:

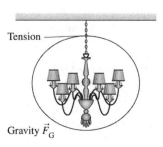

5.3. Model: Assume friction is negligible compared to other forces.
Visualize:

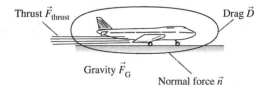

Section 5.4 What Do Forces Do? A Virtual Experiment

5.9. Visualize: Please refer to Figure EX5.9.
Solve: Newton's second law is $F = ma$. Applying this to curves 1 at the point $F = 3$ rubber bands and to curve 2 at the point $F = 5$ rubber bands gives

$$\left.\begin{array}{l} 3F = m_1(5a_1) \\ 5F = m_2(4a_1) \end{array}\right\} \frac{3}{5} = \frac{5m_1}{4m_2} \quad \Rightarrow \quad \frac{m_1}{m_2} = \frac{12}{25}$$

Assess: The line with the steepest slope should have the smallest mass, so we expect $m_1 < m_2$, which is consistent with our calculation.

5.11 Solve: Use proportional reasoning. Let T = period of the pendulum, L = length of pendulum. We are given $T \propto \sqrt{L}$, so $T/\sqrt{L}$ should be constant. We have

$$\frac{3.0 \text{ s}}{\sqrt{2.0 \text{ m}}} = \frac{x}{\sqrt{3.0 \text{ m}}} \quad \Rightarrow \quad x = \frac{\sqrt{3.0 \text{ m}}}{\sqrt{2.0 \text{ m}}}(3.0 \text{ s}) = 3.7 \text{ s}$$

Solving, the period of the 3.0 m long pendulum is $x = 3.7$ s.

Assess: Increasing the length increases the period, as expected.

Section 5.5 Newton's Second Law

5.13. Solve: Newton's second law tells us that $F = ma$. Compute F for each case:

(a) $F = (0.200 \text{ kg})(5 \text{ m/s}^2) = 1 \text{ N}$.

(b) $F = (0.200 \text{ kg})(10 \text{ m/s}^2) = 2 \text{ N}$.

Assess: To double the acceleration we must double the force, as expected.

5.17. Solve: (a) This problem calls for an *estimate* so we are looking for an approximate answer. Table 5.1 gives us no information on pencils, but does give us the weight of the U.S. quarter. Put the quarter on one hand and a pencil on the other hand. The sensation on your hand is the weight of the object. The sensation from the quarter is about the same as the sensation from the pencil, so they both have about the same weight. We can estimate the weight of the pencil to be 0.05 N.

(b) According to Table 5.1, the propulsion force on a car is 5000 N. The mass of a sprinter is about 100 kg. This is about one-tenth of the mass of a car, which is about 1000 kg for a compact model. The acceleration of a sprinter is somewhat less than that of a car, let's guess about one-fifth. We can write Newton's second law as follows:

$$F(\text{sprinter}) = \frac{1}{10}(\text{mass of car}) \times \frac{1}{5}(\text{acceleration of car}) = \frac{5000 \text{ N}}{50} = 100 \text{ N}$$

So, we would *roughly estimate* the propulsion force of a sprinter to be 100 N.

Assess: This is the same estimated number as we obtained in Exercise 5.16. This is reasonable because, in both the cases, the propulsion force comes from the human body and it the manner in which the force is delivered is not very significant.

Section 5.7 Free-Body Diagrams

5.21. Visualize:

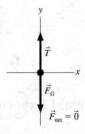

Solve: The free-body diagram shows two equal and opposite forces such that the net force is zero. The force directed down is labeled as a gravitational force, and the force directed up is labeled as a tension. With zero net force the acceleration is zero. So, a possible description is "an object hangs from a rope and is at rest" or "an object hanging from a rope is moving up or down with a constant speed."

5.23. Visualize:

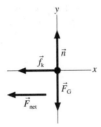

Solve: The free-body diagram shows three forces. There is a gravitational force $\vec{F}_G$, which is down. There is a normal force labeled $\vec{n}$, which is up. The forces $\vec{F}_G$ and $\vec{n}$ are shown with vectors of the same length so they are equal in magnitude and the net vertical force is zero. So we have an object on a surface and which is not moving vertically. The only horizontal force is a kinetic friction force $\vec{f}_k$ that acts to the left, so the velocity of the object must be to the right because friction always acts against the velocity. This means there is a net force to the left producing an acceleration to the left. A possible description is "a baseball player sliding into second base."

5.25. Visualize:

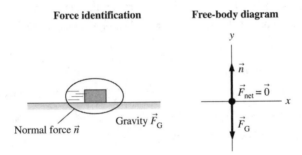

Assess: The problem says that there is no friction and it tells you nothing about any drag; so we do not include either of these forces. The only remaining forces are the weight and the normal force.

5.29. Visualize:

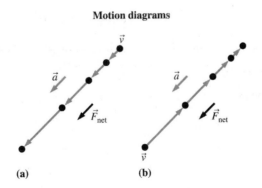

The velocity vector in figure (a) is shown downward and to the left, so movement is downward and to the left. The velocity vectors get successively longer, which means the speed is increasing. Therefore the acceleration is downward and to the left, as shown. By Newton's second law $\vec{F} = m\vec{a}$, the net force must be in the same direction as the acceleration. Thus, the net force is downward and to the left.

The velocity vector in (b) is shown to be upward and to the right. So movement is upward and to the right. The velocity vector gets successively shorter, which means the speed is decreasing. Therefore the acceleration is downward and to the left, as shown From Newton's second law, the net force must be in the direction of the acceleration, so it is directed downward and to the left.

5.31. Visualize:

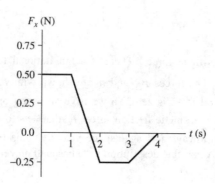

Solve: According to Newton's second law $F = ma$, the force at any time is found simply by multiplying the value of the acceleration by the mass of the object. Thus, for example, the point at (1 s, 1 m/s²) become (1 s, 1 m/s² × 0.5 kg) = (0 s, 0.5 N).

5.35. Model: Use the particle model for the object.

Solve: (a) We are told that, for an unknown force (call it F_0) acting on an unknown mass (call it m_0), the acceleration of the mass is $a_0 = 10$ m/s². According to Newton's second law, $F_0 = m_0 a_0$ so $F_0/m_0 = a_0 = 10$ m/s². If the force becomes $F' = \frac{1}{2}F_0$, Newton's second law gives

$$F' = m_0 a'$$
$$F_0/2 = m_0 a' \quad \Rightarrow \quad a' = F_0/(2m_0) = \tfrac{1}{2}a_0 = 5.0 \text{ m/s}^2$$

(b) The force is F_0 and the mass is now $m' = \frac{1}{2}m_0$. Newton's second law gives

$$F_0 = m' a'$$
$$F_0 = \tfrac{1}{2}m_0 a' \quad \Rightarrow \quad a' = 2F_0/m_0 = 2a_0 = 20 \text{ m/s}^2$$

(c) The force is $F' = \frac{1}{2}F_0$ and the mass is $m' = \frac{1}{2}m_0$. By inspection of Newton's second law, it is evident that the acceleration stays the same, so $a' = 10$ m/s².

(d) The force is $F' = \frac{1}{2}F_0$ and the mass is $m' = 2m_0$ Newton's second law gives

$$F' = m' a'$$
$$\tfrac{1}{2}F_0 = 2m_0 a' \quad \Rightarrow \quad a' = \tfrac{1}{4}F_0/m_0 = \tfrac{1}{4}a_0 = 2.5 \text{ m/s}^2$$

5.37. Visualize:

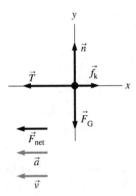

Solve: (d) There are a normal force and a gravitational force that are equal and opposite, so this is an object on a horizontal surface. The object is being pulled to the left with a nonzero net force, so it is accelerating to the left. Because the friction force is kinetic, we know that the velocity is nonzero, and is most likely in the direction of the net force. The description could be "a tow truck pulls a car out of the mud."

5.39. Visualize:

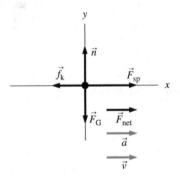

Solve: (d) This is an object on a horizontal surface because $F_G = n$. It must be moving to the right because the kinetic friction is to the left. It is experiencing a net force to the right so it is accelerating to the right. The description of the free-body diagram could be "a compressed spring is pushing a wooden block to the right over a table top."

5.43. Visualize:

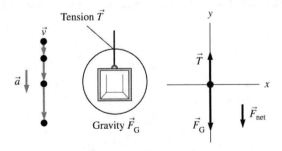

Tension is the only contact force. The downward acceleration implies that $F_G > T$.

5.47. Visualize:

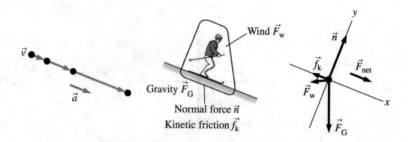

The normal force is perpendicular to the hill. The kinetic frictional force is parallel to the hill and directed upward opposite to the direction of motion. The wind force is given as *horizontal*. Since the skier stays on the slope (that is, there is no acceleration away from the slope) the net force must be parallel to the slope.

5.49. Visualize:

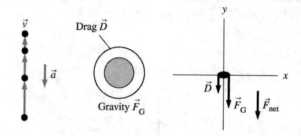

The drag force due to air is directed opposite to the motion.

5.51. Visualize:

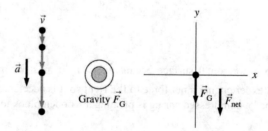

There are no contact forces on the rock. The gravitational force is the only force acting on the rock.

DYNAMICS I: MOTION ALONG A LINE

Exercises and Problems

Section 6.1 Equilibrium

6.1. Model: We can assume that the ring is a single massless particle in static equilibrium.
Visualize:

Pictorial representation

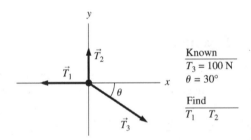

Known
$T_3 = 100$ N
$\theta = 30°$

Find
T_1 T_2

Solve: Written in component form, Newton's first law is
$$(F_{net})_x = \Sigma F_x = T_{1x} + T_{2x} + T_{3x} = 0 \text{ N} \quad (F_{net})_y = \Sigma F_y = T_{1y} + T_{2y} + T_{3y} = 0 \text{ N}$$
Evaluating the components of the force vectors from the free-body diagram:
$$T_{1x} = -T_1 \quad T_{2x} = 0 \text{ N} \quad T_{3x} = T_3 \cos 30°$$
$$T_{1y} = 0 \text{ N} \quad T_{2y} = T_2 \quad T_{3y} = -T_3 \sin 30°$$
Using Newton's first law:
$$-T_1 + T_3 \cos 30° = 0 \text{ N} \quad T_2 - T_3 \sin 30° = 0 \text{ N}$$
Rearranging:
$$T_1 = T_3 \cos 30° = (100 \text{ N})(0.8666) = 86.7 \text{ N} \quad T_2 = T_3 \sin 30° = (100 \text{ N})(0.5) = 50.0 \text{ N}$$
Assess: Since $\vec{T_3}$ acts closer to the x-axis than to the y-axis, it makes sense that $T_1 > T_2$.

6.3. Model: We assume the speaker is a particle in static equilibrium under the influence of three forces: gravity and the tensions in the two cables.

Visualize:

Pictorial representation

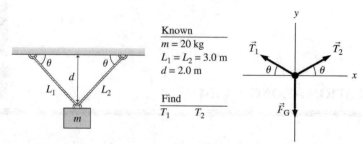

Known
m = 20 kg
$L_1 = L_2 = 3.0$ m
d = 2.0 m

Find
T_1 T_2

Solve: From the lengths of the cables and the distance below the ceiling we can calculate θ as follows:

$$\sin\theta = \frac{2 \text{ m}}{3 \text{ m}} = 0.677 \Rightarrow \theta = \sin^{-1} 0.667 = 41.8°$$

Newton's first law for this situation is

$$(F_{\text{net}})_x = \Sigma F_x = T_{1x} + T_{2x} = 0 \text{ N} \Rightarrow -T_1 \cos\theta + T_2 \cos\theta = 0 \text{ N}$$
$$(F_{\text{net}})_y = \Sigma F_y = T_{1y} + T_{2y} + w_y = 0 \text{ N} \Rightarrow T_1 \sin\theta + T_2 \sin\theta - w = 0 \text{ N}$$

The x-component equation means $T_1 = T_2$. From the y-component equation:

$$2T_1 \sin\theta = w \Rightarrow T_1 = \frac{w}{2\sin\theta} = \frac{mg}{2\sin\theta} = \frac{(20 \text{ kg})(9.8 \text{ m/s}^2)}{2\sin 41.8°} = \frac{196 \text{ N}}{1.333} = 147 \text{ N}$$

Assess: It's to be expected that the two tensions are equal, since the speaker is suspended symmetrically from the two cables. That the two tensions add to considerably more than the weight of the speaker reflects the relatively large angle of suspension.

Section 6.2 Using Newton's Second Law

6.7. Solve: (a) For the diagram on the left, three of the vectors lie along the axes of the tilted coordinate system. Notice that the angle between the 3 N force and the $-y$-axis is the same 20° by which the coordinates are tilted. Applying Newton's second law,

$$a_x = \frac{(F_{\text{net}})_x}{m} = \frac{5.0 \text{ N} - 1.0 \text{ N} - (3.0\sin 20°) \text{ N}}{2.0 \text{ kg}} = 1.49 \text{ m/s}^2 \approx 1.5 \text{ m/s}^2$$

$$a_y = \frac{(F_{\text{net}})_y}{m} = \frac{2.82 \text{ N} - (3.0\cos 20°) \text{ N}}{2.0 \text{ kg}} = 0 \text{ m/s}^2$$

(b) For the diagram on the right, the 2-newton force in the first quadrant makes an angle of 15° with the positive x-axis. The other 2-newton force makes an angle of 15° with the negative y-axis. The accelerations are

$$a_x = \frac{(F_{\text{net}})_x}{m} = \frac{(2.0\cos 15°) \text{ N} + (2.0\sin 15°) \text{ N} - 3.0 \text{ N}}{2.0 \text{ kg}} = -0.28 \text{ m/s}^2$$

$$a_y = \frac{(F_{\text{net}})_y}{m} = \frac{1.414 \text{ N} + (2.0\sin 15°) \text{ N} - (2.0\cos 15°) \text{ N}}{2.0 \text{ kg}} = 0 \text{ m/s}^2$$

6.9. Visualize: Assuming the positive direction is to the right, positive forces result in the object accelerating to the right and negative forces result in the object accelerating to the left. The final segment of zero force is a period of constant speed.
Solve: We have the mass and net force for all the three segments. This means we can use Newton's second law to calculate the accelerations. The acceleration from $t = 0$ s to $t = 3$ s is

$$a_x = \frac{F_x}{m} = \frac{4 \text{ N}}{2.0 \text{ kg}} = 2 \text{ m/s}^2$$

The acceleration from $t = 3$ s to $t = 5$ s is

$$a_x = \frac{F_x}{m} = \frac{-2 \text{ N}}{2.0 \text{ kg}} = -1 \text{ m/s}^2$$

The acceleration from $t = 5$ s to 8 s is $a_x = 0$ m/s^2. In particular, a_x(at $t = 6$ s) = 0 m/s^2.
We can now use one-dimensional kinematics to calculate v at $t = 6$ s as follows:

$$v = v_0 + a_1(t_1 - t_0) + a_2(t_2 - t_0)$$
$$= 0 + (2 \text{ m/s}^2)(3 \text{ s}) + (-1 \text{ m/s}^2)(2 \text{ s}) = 6 \text{ m/s} - 2 \text{ m/s} = 4 \text{ m/s}$$

Assess: The positive final velocity makes sense, given the greater magnitude and longer duration of the positive $\vec{F}_1$. A velocity of 4 m/s also seems reasonable, given the magnitudes and directions of the forces and the mass involved.

Section 6.3 Mass, Weight, and Gravity

6.13. Model: Use the particle model for the woman.
Solve: (a) The woman's weight on the earth is

$$w_{\text{earth}} = mg_{\text{earth}} = (55 \text{ kg})(9.80 \text{ m/s}^2) = 540 \text{ N}$$

(b) Since mass is a measure of the amount of matter, the woman's mass is the same on Mars as on the earth. Her weight on Mars is

$$w_{\text{Mars}} = mg_{\text{Mars}} = (55 \text{ kg})(3.76 \text{ m/s}^2) = 210 \text{ N}$$

Assess: The smaller acceleration due to gravity on Mars reveals that objects are less strongly attracted to Mars than to the earth. Thus the woman's smaller weight on Mars makes sense.

6.15. Model: We assume that the passenger is a particle acted on by only two vertical forces: the downward pull of gravity and the upward force of the elevator floor.
Visualize: The graph has three segments corresponding to different conditions: (1) increasing velocity, meaning an upward acceleration; (2) a period of constant upward velocity; and (3) decreasing velocity, indicating a period of deceleration (negative acceleration).
Solve: Given the assumptions of our model, we can calculate the acceleration for each segment of the graph and then apply Equation 6.10. The acceleration for the first segment is

$$a_y = \frac{v_1 - v_0}{t_1 - t_0} = \frac{8 \text{ m/s} - 0 \text{ m/s}}{2 \text{ s} - 0 \text{ s}} = 4 \text{ m/s}^2$$

$$\Rightarrow w = mg\left(1 + \frac{a_y}{g}\right) = mg\left(1 + \frac{4 \text{ m/s}^2}{9.80 \text{ m/s}^2}\right) = (75 \text{ kg})(9.80 \text{ m/s}^2)\left(1 + \frac{4}{9.80}\right) = 1035 \text{ N}$$

For the second segment, $a_y = 0$ m/s^2 and the weight is

$$w = mg\left(1 + \frac{0 \text{ m/s}^2}{g}\right) = mg = (75 \text{ kg})(9.80 \text{ m/s}^2) = 740 \text{ N}$$

For the third segment,

$$a_y = \frac{v_3 - v_2}{t_3 - t_2} = \frac{0 \text{ m/s} - 8 \text{ m/s}}{10 \text{ s} - 6 \text{ s}} = -2 \text{ m/s}^2$$

$$\Rightarrow w = mg\left(1 + \frac{-2 \text{ m/s}^2}{9.80 \text{ m/s}^2}\right) = (75 \text{ kg})(9.80 \text{ m/s}^2)(1 - 0.2) = 590 \text{ N}$$

Assess: As expected, the weight is greater than the gravitational force on the passenger when the elevator is accelerating upward and lower than normal when the acceleration is downward. When there is no acceleration the weight is the gravitational force. In all three cases the magnitudes are reasonable, given the mass of the passenger and the accelerations of the elevator.

Section 6.4 Friction

6.19. Model: We will represent the crate as a particle.

Visualize:

Pictorial representation

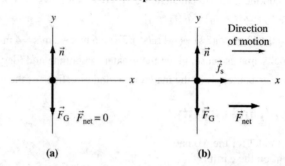

(a)   (b)

Solve: (a) When the belt runs at constant speed, the crate has an acceleration $\vec{a} = \vec{0}$ m/s² and is in dynamic equilibrium. Thus $\vec{F}_{net} = \vec{0}$. It is tempting to think that the belt exerts a friction force on the crate. But if it did, there would be a *net* force because there are no other possible horizontal forces to balance a friction force. Because there is no net force, there cannot be a friction force. The only forces are the upward normal force and the gravitational force on the crate. (A friction force would have been needed to get the crate moving initially, but no horizontal force is needed to keep it moving once it is moving with the same constant speed as the belt.)
(b) If the belt accelerates gently, the crate speeds up without slipping on the belt. Because it is accelerating, the crate must have a net horizontal force. So *now* there is a friction force, and the force points in the direction of the crate's motion. Is it static friction or kinetic friction? Although the crate is moving, there is *no* motion of the crate relative to the belt. Thus, it is a *static* friction force that accelerates the crate so that it moves without slipping on the belt.
(c) The static friction force has a maximum possible value $(f_s)_{max} = \mu_s n$. The maximum possible acceleration of the crate is

$$a_{max} = \frac{(f_s)_{max}}{m} = \frac{\mu_s n}{m}$$

If the belt accelerates more rapidly than this, the crate will not be able to keep up and will slip. It is clear from the free-body diagram that $n = F_G = mg$. Thus,

$$a_{max} = \mu_s g = (0.5)(9.80 \text{ m/s}^2) = 4.9 \text{ m/s}^2$$

6.21. Model: We assume that the truck is a particle in equilibrium, and use the model of static friction.
Visualize:

Pictorial representation

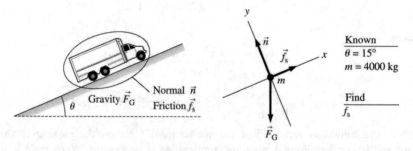

Known
$\theta = 15°$
$m = 4000$ kg

Find
f_s

Solve: The truck is not accelerating, so it is in equilibrium, and we can apply Newton's first law. The normal force has no component in the x-direction, so we can ignore it here. For the other two forces:

$$(F_{net})_x = \Sigma F_x = f_s - (F_G)_x = 0 \text{ N} \Rightarrow f_s = (F_G)_x = mg\sin\theta = (4000 \text{ kg})(9.8 \text{ m/s}^2)(\sin 15°) = 10,145 \text{ N} \approx 10,000 \text{ N}$$

Assess: The truck's weight (mg) is roughly 40,000 N. A friction force that is $\approx 25\%$ of the truck's weight seems reasonable.

Section 6.5 Drag

6.25. Model: We will represent the tennis ball as a particle. The drag coefficient is 0.5.

Visualize:

Pictorial representation

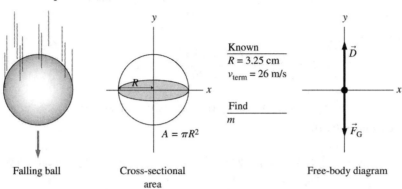

Falling ball	Cross-sectional area	Free-body diagram

The tennis ball falls straight down toward the earth's surface. The ball is subject to a net force that is the resultant of the gravitational and drag force vectors acting vertically, in the downward and upward directions, respectively. Once the net force acting on the ball becomes zero, the terminal velocity is reached and remains constant for the rest of the motion.

Solve: The mathematical equation defining the dynamical equilibrium situation for the falling ball is
$$\vec{F}_{net} = \vec{F}_G + \vec{D} = \vec{0} \text{ N}$$
Since only the vertical direction matters, one can write:
$$\Sigma F_y = 0 \text{ N} \Rightarrow F_{net} = D - F_G = 0 \text{ N}$$
When this condition is satisfied, the speed of the ball becomes the constant terminal speed $v = v_{term}$. The magnitudes of the gravitational and drag forces acting on the ball are:
$$F_G = mg = m(9.80 \text{ m/s}^2)$$
$$D \approx \frac{1}{2}(C\rho A v_{term}^2) = 0.5(0.5)(1.2 \text{ kg/m}^3)(\pi R^2)v_{term}^2 = (0.3\pi)(0.0325 \text{ m})^2(26 \text{ m/s})^2 = 0.67 \text{ N}$$
The condition for dynamic equilibrium becomes:
$$(9.80 \text{ m/s}^2)m - 0.67 \text{ N} = 0 \text{ N} \Rightarrow m = \frac{0.67 \text{ N}}{9.80 \text{ m/s}^2} = 69 \text{ g}$$

Assess: The value of the mass of the tennis ball obtained above seems reasonable.

6.29. Model: The plastic ball is represented as a particle in static equilibrium.
Visualize:

Pictorial representation

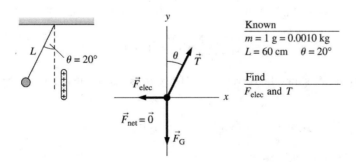

Solve: (a) The electric force, like the weight, is a long-range force. So the ball experiences the contact force of the string's tension plus *two* long-range forces. The equilibrium condition is

$$(F_{net})_x = T_x + (F_{elec})_x = T\sin\theta - F_{elec} = 0 \text{ N}$$
$$(F_{net})_y = T_y + (F_G)_y = T\cos\theta - mg = 0 \text{ N}$$

We can solve the *y*-equation to get

$$T = \frac{mg}{\cos\theta} = \frac{(0.001 \text{ kg})(9.8 \text{ m/s}^2)}{\cos 20°} = 0.0104 \text{ N}$$

Substituting this value into the *x*-equation,

$$F_{elec} = T\sin\theta = (1.04\times10^{-2} \text{ N})\sin 20° = 0.0036 \text{ N}$$

(b) The tension in the string is $0.0104 \text{ N} \approx 0.010 \text{ N}$.

6.31. Model: We will represent Henry as a particle. His motion is governed by constant-acceleration kinematic equations.

Solve: (a) Henry undergoes an acceleration from 0 s to 2.0 s, constant velocity motion from 2.0 s to 10.0 s, and another acceleration as the elevator brakes from 10.0 s to 12.0 s. The weight is the same as the gravitational force during constant velocity motion, so Henry's weight $w = F_G = mg$ is 750 N. His weight is *less* than the gravitational force on him during the initial acceleration, so the acceleration is in a *downward* direction (negative *a*). Thus, the elevator's initial motion is *down*.

(b) Because the gravitational force on Henry is 750 N, his mass is $m = F_G/g = 76.5 \text{ kg} \approx 77 \text{ kg}$.

(c) The apparent weight during vertical motion is given by

$$w = mg\left(1+\frac{a}{g}\right) \Rightarrow a = g\left(\frac{w}{F_G}-1\right)$$

During the interval $0 \text{ s} \leq t \leq 2 \text{ s}$, the elevator's acceleration is

$$a = g\left(\frac{600 \text{ N}}{750 \text{ N}}-1\right) = -1.96 \text{ m/s}^2$$

At $t = 2$ s, Henry's position is

$$y_1 = y_0 + v_0\Delta t_0 + \frac{1}{2}a(\Delta t_0)^2 = \frac{1}{2}a(\Delta t_0)^2 = -3.92 \text{ m}$$

and his velocity is

$$v_1 = v_0 + a\Delta t_0 = a\Delta t_0 = -3.92 \text{ m/s}$$

During the interval $2 \text{ s} \leq t \leq 10 \text{ s}$, $a = 0 \text{ m/s}^2$. This means Henry travels with a constant velocity $v_1 = -3.92$ m/s. At $t = 10$ s he is at position

$$y_2 = y_1 + v_1\Delta t_1 = -35.3 \text{ m}$$

and he has a velocity $v_2 = v_1 = -3.92$ m/s. During the interval $10 \text{ s} \leq t \leq 12.0 \text{ s}$, the elevator's acceleration is

$$a = g\left(\frac{900 \text{ N}}{750 \text{ N}}-1\right) = +1.96 \text{ m/s}^2$$

The upward acceleration vector slows the elevator and Henry feels heavier than normal. At $t = 12.0$ s Henry is at position

$$y_3 = y_2 + v_2(\Delta t_2) + \frac{1}{2}a(\Delta t_2)^2 = -39.2 \text{ m}$$

Thus Henry has traveled distance $39.2 \text{ m} \approx 39 \text{ m}$.

6.33. Model: We can assume the foot is a single particle in equilibrium under the combined effects of gravity, the tensions in the upper and lower sections of the traction rope, and the opposing traction force of the leg itself. We can also treat the hanging mass as a particle in equilibrium. Since the pulleys are frictionless, the tension is the same everywhere in the rope. Because all pulleys are in equilibrium, their net force is zero. So they do not contribute to T.
Visualize:

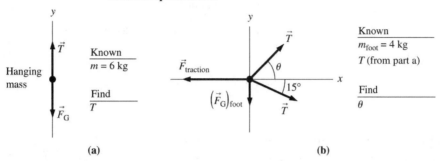

Solve: (a) From the free-body diagram for the mass, the tension in the rope is
$$T = F_G = mg = (6 \text{ kg})(9.80 \text{ m/s}^2) = 58.8 \text{ N} \approx 59 \text{ N}$$
(b) Using Newton's first law for the vertical direction on the pulley attached to the foot,
$$(F_{net})_y = \Sigma F_y = T\sin\theta - T\sin 15° - (F_G)_{foot} = 0 \text{ N}$$
$$\Rightarrow \sin\theta = \frac{T\sin 15° + (F_G)_{foot}}{T} = \sin 15° + \frac{m_{foot}g}{T} = 0.259 + \frac{(4 \text{ kg})(9.80 \text{ m/s}^2)}{58.8 \text{ N}} = 0.259 + 0.667 = 0.926$$
$$\Rightarrow \theta = \sin^{-1} 0.926 = 67.8° \approx 68°$$
(c) Using Newton's first law for the horizontal direction,
$$(F_{net})_x = \Sigma F_x = T\cos\theta + T\cos 15° - F_{traction} = 0 \text{ N}$$
$$\Rightarrow F_{traction} = T\cos\theta + T\cos 15° = T(\cos 67.8° + \cos 15°)$$
$$= (58.8 \text{ N})(0.3778 + 0.9659) = (58.8 \text{ N})(1.344) = 79 \text{ N}$$

Assess: Since the tension in the upper segment of the rope must support the foot and counteract the downward pull of the lower segment of the rope, it makes sense that its angle is larger (a more direct upward pull). The magnitude of the traction force, roughly one-tenth of the gravitational force on a human body, seems reasonable.

6.37. Model: The ball is represented as a particle that obeys constant-acceleration kinematic equations.
Visualize:

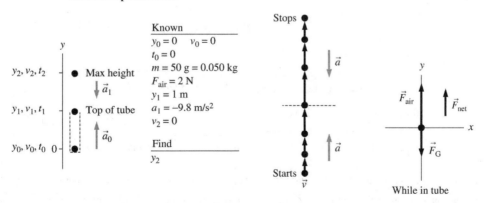

Solve: This is a two-part problem. During part 1 the ball accelerates upward in the tube. During part 2 the ball undergoes free fall $(a = -g)$. The initial velocity for part 2 is the final velocity of part 1, as the ball emerges from the tube. The free-body diagram for part 1 shows two forces: the air pressure force and the gravitational force. We need only the y-component of Newton's second law:

$$a_y = a = \frac{(F_{net})_y}{m} = \frac{F_{air} - F_G}{m} = \frac{F_{air}}{m} - g = \frac{2\ N}{0.05\ kg} - 9.80\ m/s^2 = 30.2\ m/s^2$$

We can use kinematics to find the velocity v_1 as the ball leaves the tube:

$$v_1^2 = v_0^2 + 2a(y_1 - y_0) \Rightarrow v_1 = \sqrt{2ay_1} = \sqrt{2(30.2\ m/s^2)(1\ m)} = 7.77\ m/s$$

For part 2, free-fall kinematics $v_2^2 = v_1^2 - 2g(y_2 - y_1)$ gives

$$y_2 - y_1 = \frac{v_1^2}{2g} = 3.1\ m$$

6.41. Model: Model the object as a particle. Neglect air resistance.

Visualize: We'll use $v_1^2 = v_0^2 + 2a\Delta x$ to find the speed of the object. Since $v_0 = 0$, $v_1 = \sqrt{2a_x L}$.

We'll also use Newton's second law in both directions in order to find a_x.

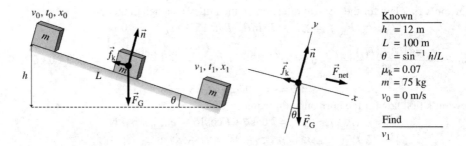

Known
$h = 12$ m
$L = 100$ m
$\theta = \sin^{-1} h/L$
$\mu_k = 0.07$
$m = 75$ kg
$v_0 = 0$ m/s

Find
v_1

Solve:
(a)

$$\Sigma F_y = n - mg\cos\theta = 0 \Rightarrow n = mg\cos\theta$$

$$\Sigma F_x = mg\sin\theta - f_k = ma_x$$

$$mg\sin\theta - \mu_k n = ma_x$$

$$mg\sin\theta - \mu_k mg\cos\theta = ma_x$$

Cancel the m.

$$a_x = g(\sin\theta - \mu_k\cos\theta)$$

Now put this back in to the equation for v_1.

$$v_1 = \sqrt{2a_x L} = \sqrt{2[g(\sin\theta - \mu_k\cos\theta)]L} = \sqrt{2g(h - \mu_k\sqrt{L^2 - h^2})}$$

(b) For $h = 12$ m, $L = 100$ m, $\mu_k = 0.07$ we have

$$v_1 = \sqrt{2(9.8\ m/s^2)\left((12\ m - 0.07)\sqrt{(100\ m)^2 - (12\ m)^2}\right)} = 9.949\ m/s \approx 9.9\ m/s$$

Assess: Sam's mass was extra unneeded information because m cancels out of the equation for v_1. Any skier, regardless of their mass, would achieve the same speed at the bottom of the same hill with the same μ_k.

6.43. Model: We assume Sam is a particle moving in a straight line down the slope under the influence of gravity, the thrust of his jet skis, and the resisting force of friction on the snow.

Visualize:

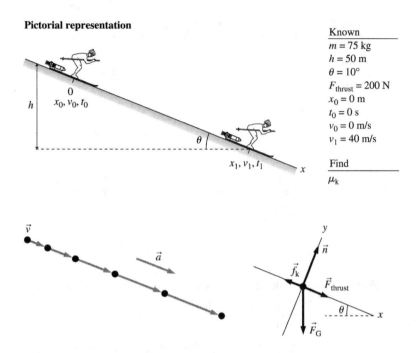

Pictorial representation

Known
$m = 75$ kg
$h = 50$ m
$\theta = 10°$
$F_{\text{thrust}} = 200$ N
$x_0 = 0$ m
$t_0 = 0$ s
$v_0 = 0$ m/s
$v_1 = 40$ m/s

Find
μ_k

Solve: From the height of the slope and its angle, we can calculate its length:

$$\frac{h}{x_1 - x_0} = \sin\theta \Rightarrow x_1 - x_0 = \frac{h}{\sin\theta} = \frac{50 \text{ m}}{\sin 10°} = 288 \text{ m}$$

Since Sam is not accelerating in the y-direction, we can use Newton's first law to calculate the normal force:

$$(F_{\text{net}})_y = \Sigma F_y = n - F_G \cos\theta = 0 \text{ N} \Rightarrow n = F_G \cos\theta = mg\cos\theta = (75 \text{ kg})(9.80 \text{ m/s}^2)(\cos 10°) = 724 \text{ N}$$

One-dimensional kinematics gives us Sam's acceleration:

$$v_1^2 = v_0^2 + 2a_x(x - x_0) \Rightarrow a_x = \frac{v_1^2 - v_0^2}{2(x_1 - x_2)} = \frac{(40 \text{ m/s})^2 - 0 \text{ m}^2/\text{s}^2}{2(288 \text{ m})} = 2.78 \text{ m/s}^2$$

Then, from Newton's second law and the equation $f_k = \mu_k n$:

$$(F_{\text{net}})_x = \Sigma F_x = F_G \sin\theta + F_{\text{thrust}} - f_k = ma_x$$

$$\Rightarrow \mu_k = \frac{mg\sin\theta + F_{\text{thrust}} - ma}{n} = \frac{(75 \text{ kg})(9.80 \text{ m/s}^2)(\sin 10°) + 200 \text{ N} - (75 \text{ kg})(2.78 \text{ m/s}^2)}{724 \text{ N}} = 0.165$$

Assess: This coefficient seems a bit high for skis on snow, but not impossible.

6.47. Model: We assume that the plane is a particle accelerating in a straight line under the influence of two forces: the thrust of its engines and the rolling friction of the wheels on the runway. We can use one-dimensional kinematics.

Visualize:

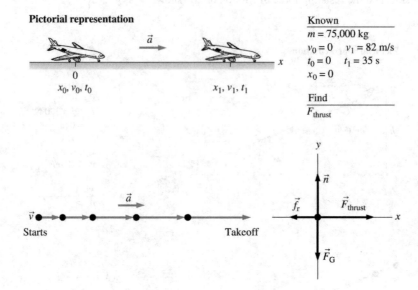

Pictorial representation

Known
$m = 75,000$ kg
$v_0 = 0$ $v_1 = 82$ m/s
$t_0 = 0$ $t_1 = 35$ s
$x_0 = 0$

Find
F_{thrust}

Solve: We can use the definition of acceleration to find a, and then apply Newton's second law. We obtain:

$$a = \frac{\Delta v}{\Delta t} = \frac{82 \text{ m/s} - 0 \text{ m/s}}{35 \text{ s}} = 2.34 \text{ m/s}^2$$

$$(F_{net}) = \Sigma F_x = F_{thrust} - f_r = ma \Rightarrow F_{thrust} = f_r + ma$$

For rubber rolling on concrete, $\mu_r = 0.02$ (Table 6.1), and since the runway is horizontal, $n = F_G = mg$. Thus:

$$F_{thrust} = \mu_r F_G + ma = \mu_r mg + ma = m(\mu_r g + a)$$
$$= (75,000 \text{ kg})[(0.02)(9.8 \text{ m/s}^2) + 2.34 \text{ m/s}^2] = 190,000 \text{ N}$$

Assess: It's hard to evaluate such an enormous thrust, but comparison with the plane's mass suggests that 190,000 N is enough to produce the required acceleration.

6.49. Model: We will model the sled and friend as a particle, and use the model of kinetic friction because the sled is in motion.
Visualize:

Pictorial representation

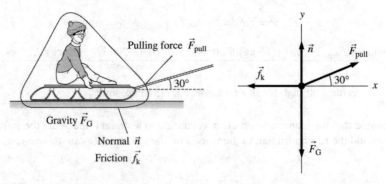

Pulling force $\vec{F}_{pull}$

Gravity $\vec{F}_G$

Normal $\vec{n}$

Friction $\vec{f}_k$

The net force on the sled is zero (note the constant speed of the sled). That means the component of the pulling force along the +x-direction is equal to the magnitude of the kinetic force of friction in the −x-direction. Also note that $(F_{net})_y = 0$ N, since the sled is not moving along the y-axis.

Solve: Newton's second law is

$$(F_{net})_x = \Sigma F_x = n_x + (F_G)_x + (f_k)_x + (F_{pull})_x = 0\ \text{N} + 0\ \text{N} - f_k + F_{pull}\cos\theta = 0\ \text{N}$$

$$(F_{net})_y = \Sigma F_y = n_y + (F_G)_y + (f_k)_y + (F_{pull})_y = n - mg + 0\ \text{N} + F_{pull}\sin\theta = 0\ \text{N}$$

The x-component equation using the kinetic friction model $f_k = \mu_k n$ reduces to

$$\mu_k n = F_{pull}\cos\theta$$

The y-component equation gives

$$n = mg - F_{pull}\sin\theta$$

We see that the normal force is smaller than the gravitational force because F_{pull} has a component in a direction opposite to the direction of the gravitational force. In other words, F_{pull} is partly lifting the sled. From the x-component equation, μ_k can now be obtained as

$$\mu_k = \frac{F_{pull}\cos\theta}{mg - F_{pull}\sin\theta} = \frac{(75\ \text{N})(\cos 30°)}{(60\ \text{kg})(9.80\ \text{m/s}^2) - (75\ \text{N})(\sin 30°)} = 0.12$$

Assess: A quick glance at the various μ_k values in Table 6.1 suggests that a value of 0.12 for μ_k is reasonable.

6.53. Model: The antiques (mass $= m$) in the back of your pickup (mass $= M$) will be treated as a particle. The antiques touch the truck's steel bed, so only the steel bed can exert contact forces on the antiques. The pickup-antiques system will also be treated as a particle, and the contact force on this particle will be due to the road.
Visualize:

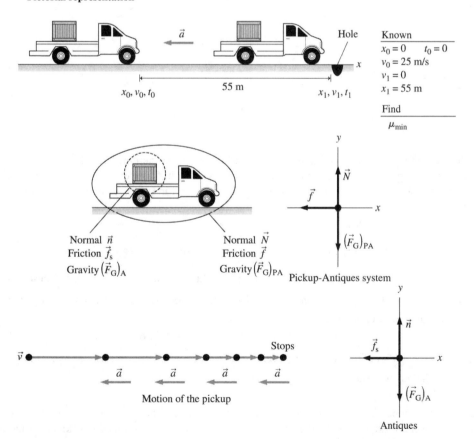

Solve: (a) We will find the smallest coefficient of friction that allows the truck to stop in 55 m, then compare that to the known coefficients for rubber on concrete. For the pickup-antiques system, with mass $m + M$, Newton's second law is

$$(F_{net})_x = \sum F_x = N_x + ((F_G)_{PA})_x + (f)_x = 0\ N + 0\ N - f = (m+M)a_x = (m+M)a$$

$$(F_{net})_y = \sum F_y = N_y + ((F_G)_{PA})_y + (f)_y = N - (m+M)g + 0\ N = 0\ N$$

The model of static friction is $f = \mu N$, where μ is the coefficient of friction between the tires and the road. These equations can be combined to yield $a = -\mu g$. Since constant-acceleration kinematics gives $v_1^2 = v_0^2 + 2a(x_1 + x_0)$, we find

$$a = \frac{v_1^2 - v_0^2}{2(x_1 - x_0)} \Rightarrow \mu_{min} = \frac{v_0^2}{2g(x_1 - x_0)} = \frac{(25\ \text{m/s})^2}{(2)(9.8\ \text{m/s}^2)(55\ \text{m})} = 0.58$$

The truck cannot stop if μ is smaller than this. But both the static and kinetic coefficients of friction, 1.00 and 0.80 respectively (see Table 6.1), are larger. So the truck can stop.

(b) The analysis of the pickup-antiques system applies to the antiques, and it gives the same value of 0.58 for μ_{min}. This value is smaller than the given coefficient of static friction ($\mu_s = 0.60$) between the antiques and the truck bed. Therefore, the antiques will not slide as the truck is stopped over a distance of 55 m.

Assess: The analysis of parts (a) and (b) are the same because mass cancels out of the calculations. According to the California Highway Patrol Web site, the stopping distance (with zero reaction time) for a passenger vehicle traveling at 25 m/s or 82 ft/s is approximately 43 m. This is smaller than the 55 m over which you are asked to stop the truck.

6.55. Model: Use the particle model for the block and the model of static friction.
Visualize:

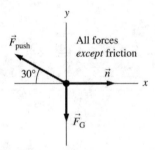

Solve: The block is initially at rest, so initially the friction force is static friction. If the 12 N push is too strong, the box will begin to move up the wall. If it is too weak, the box will begin to slide down the wall. And if the pushing force is within the proper range, the box will remain stuck in place. First, let's evaluate the sum of all the forces *except* friction:

$$\sum F_x = n - F_{push} \cos 30° = 0\ N \Rightarrow n = F_{push} \cos 30°$$

$$\sum F_y = F_{push} \sin 30° - F_G = F_{push} \sin 30° - mg = (12\ \text{N}) \sin 30° - (1\ \text{kg})(9.8\ \text{m/s}^2) = -3.8\ \text{N}$$

In the first equation we utilized the fact that any motion is parallel to the wall, so $a_x = 0\ \text{m/s}^2$. These three forces add up to $-3.8\hat{j}$ N. This means the static friction force will be able to prevent the box from moving if $f_s = +3.8\hat{j}$ N. Using the x-equation and the friction model we get

$$(f_s)_{max} = \mu_s n = \mu_s F_{push} \cos 30° = 5.2\ \text{N}$$

where we used $\mu_s = 0.5$ for wood on wood. The static friction force $\vec{f_s}$ needed to keep the box from moving is *less* than $(f_s)_{max}$. Thus the box will stay at rest.

6.57. Model: We will model the skier along with the wooden skis as a particle of mass m. The snow exerts a contact force and the wind exerts a drag force on the skier. We will therefore use the models of kinetic friction and drag. Assume the skier is a cylinder end-forward so that $C = 0.8$.

Visualize:

Pictorial representation

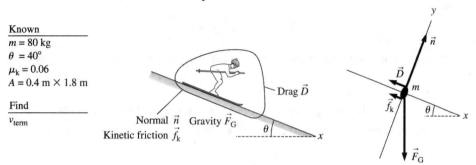

Known
$m = 80$ kg
$\theta = 40°$
$\mu_k = 0.06$
$A = 0.4$ m $\times$ 1.8 m

Find
v_{term}

Normal $\vec{n}$ Gravity $\vec{F}_G$
Kinetic friction $\vec{f}_k$

Drag $\vec{D}$

We choose a coordinate system such that the skier's motion is along the $+x$-direction. While the forces of kinetic friction $\vec{f}_k$ and drag $\vec{D}$ act along the $-x$-direction opposing the motion of the skier, the gravitational force on the skier has a component in the $+x$-direction. At the terminal speed, the net force on the skier is zero as the forces along the $+x$-direction cancel out the forces along the $-x$-direction.

Solve: Newton's second law and the models of kinetic friction and drag are

$$(F_{net})_x = \Sigma F_x = +(F_G)_x + (f_k)_x + (D)_x = mg\sin\theta - f_k - \frac{1}{2}C\rho A v^2 = ma_x = 0 \text{ N}$$

$$(F_{net})_y = \Sigma F_y = n_y + (F_G)_y = n - mg\cos\theta = 0 \text{ N}$$

$$f_k = \mu_k n$$

These three equations can be combined together as follows:

$$(1/2)C\rho A v^2 = mg\sin\theta - f_k = mg\sin\theta - \mu_k n = mg\sin\theta - \mu_k\, mg\cos\theta$$

$$\Rightarrow v_{term} = \left(mg\frac{\sin\theta - \mu_k\cos\theta}{\frac{1}{2}C\rho A}\right)^{1/2}$$

Using $\mu_k = 0.06$ and $A = 1.8$ m$\times 0.40$ m $= 0.72$ m^2, we find

$$v_{term} = \left[(80 \text{ kg})(9.8 \text{ m/s}^2)\left(\frac{\sin 40° - 0.06\cos 40°}{\frac{1}{2}(0.8)(1.2 \text{ kg/m}^3)(0.72 \text{ m}^2)}\right)\right]^{1/2} = 37 \text{ m/s}$$

Assess: A terminal speed of 37 m/s corresponds to a speed of ≈ 82 mph. This speed is reasonable but high due to the steep slope angle of 40° and a small coefficient of friction.

6.63. Model: Model the object as a particle. The acceleration is not constant so we can't use the kinematic equations. All the motion is in the x-direction.

Visualize: Divide F by m to get a and then integrate twice. The constants of integration are both zero because of the initial conditions.

Solve:

$$a_x(t) = \frac{F_x}{m} = \frac{F_0}{m}\left(1 - \frac{t}{T}\right)$$

(a)

$$v_x(t) = \int a_x\, dt = \frac{F_0}{m}\int\left(1 - \frac{t}{T}\right)dt = \frac{F_0}{m}\left(t - \frac{t^2}{2T}\right) + v_0 = \frac{F_0}{m}\left(t - \frac{t^2}{2T}\right)$$

$$v_x(T) = \frac{F_0}{m}\left(T - \frac{T^2}{2T}\right) = \frac{F_0\,T}{m\,2}$$

(b)

$$x(t) = \int v_x \, dt = \frac{F_0}{m} \int \left(t - \frac{t^2}{2T} \right) dt = \frac{F_0}{m} \left(\frac{t^2}{2} - \frac{t^3}{6T} \right) + x_0 = \frac{F_0}{m} \left(\frac{t^2}{2} - \frac{t^3}{6T} \right)$$

$$x(T) = \frac{F_0}{m} \left(\frac{T^2}{2} - \frac{T^3}{6T} \right) = \frac{F_0}{m} \frac{T^2}{3}$$

Assess: It seems reasonable that the velocity after time T would increase with T and that the position at time T would increase with T^2.

6.65. Model: Use the linear model of drag. Assume the microorganisms are swimming in water at $20°C$.
Visualize: The viscosity of water is $\eta = 1.0 \times 10^{-3}$ N·s/m² at $20°C$.
Solve:
(a)

$$\sum \vec{F} = \vec{F}_{prop} - \vec{D} = 0 \Rightarrow F_{prop} = 6\pi \eta R v$$

For a paramecium

$$F_{prop} = 6\pi (1.0 \times 10^{-3} \text{ N·s/m}^2)(50 \times 10^{-6} \text{ m})(0.0010 \text{ m/s}) = 9.4 \times 10^{-10} \text{ N}$$

For an *E. coli* bacterium

$$F_{prop} = 6\pi (1.0 \times 10^{-3} \text{ N·s/m}^2)(1.0 \times 10^{-6} \text{ m})(30 \times 10^{-6} \text{ m/s}) = 5.7 \times 10^{-13} \text{ N}$$

(b)

$$a = \frac{F_{prop}}{m} = \frac{F_{prop}}{\rho V} = \frac{F_{prop}}{\rho \frac{4}{3} \pi R^2}$$

For a paramecium

$$a = \frac{9.4 \times 10^{-10} \text{ N}}{(1000 \text{ kg/m}^3) \frac{4}{3} \pi (50 \times 10^{-6} \text{ m})^3} = 1.8 \text{ m/s}^2$$

For an *E. coli* bacterium

$$a = \frac{5.7 \times 10^{-13} \text{ N}}{(1000 \text{ kg/m}^3) \frac{4}{3} \pi (1.0 \times 10^{-6} \text{ m})^3} = 135 \text{ m/s}^2$$

Assess: The two accelerations are within a factor of two of each other.

7

NEWTON'S THIRD LAW

Exercises and Problems

Section 7.2 Analyzing Interacting Objects

7.1. Visualize:

Sketch

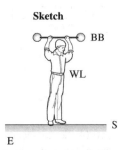

Solve: (a) The weight lifter is holding the barbell in dynamic equilibrium as he stands up, so the net force on the barbell and on the weight lifter must be zero. The barbells have an upward contact force from the weight lifter and the gravitational force downward. The weight lifter has a downward contact force from the barbells and an upward one from the surface. Gravity also acts on the weight lifter.

Interaction diagram

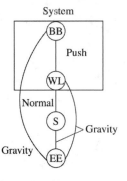

BB = Barbells
WL = Weight lifter
S = Surface EE = Entire Earth

(b) The system is the weight lifter and barbell, as indicated in the figure.

(c)

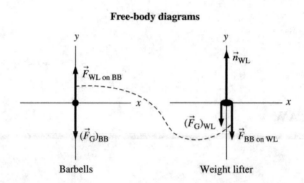

Free-body diagrams

Barbells   Weight lifter

7.5. Visualize: Please refer to Figure EX7.5.

Solve: (a) Gravity acts on both blocks. Block A is in contact with the floor and experiences a normal force and friction. The string tension is the same on both blocks since the rope and pulley are massless and the pulley is frictionless. There are two third law pair of forces at the surface where the two blocks touch. Block B pushes against Block A with a normal force, while Block A pushes back against Block B. There is also friction between the two blocks at the surface.

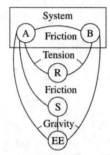

Interaction diagram

(b) A string that will not stretch constrains the two blocks to accelerate at the same rate but in opposite directions. Block A accelerates down the incline with an acceleration equal in magnitude to the acceleration of Block B up the incline. The system consists of the two blocks, as indicated in the figure above.

(c)

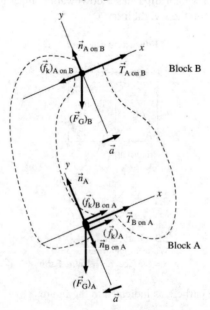

Free-body diagrams

Assess: The inclined coordinate systems allows the acceleration a to be purely along the x-axis. This is convenient because it simplifyies the mathematical expression of Newton's second law.

Section 7.3 Newton's Third Law

7.7. Model: We will model the astronaut and the chair as particles. The astronaut and the chair will be denoted by A and C, respectively, and they are separate systems. The launch pad is a part of the environment.
Visualize:

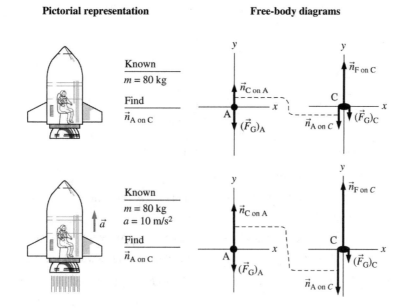

Solve: **(a)** Newton's second law for the astronaut is

$$\Sigma(F_{\text{on A}})_y = n_{\text{C on A}} - (F_G)_A = m_A a_A = 0 \text{ N} \quad \Rightarrow \quad n_{\text{C on A}} = (F_G)_A = m_A g$$

By Newton's third law, the astronaut's force on the chair is

$$n_{\text{A on C}} = n_{\text{C on A}} = m_A g = (80 \text{ kg})(9.8 \text{ m/s}^2) = 7.8 \times 10^2 \text{ N}$$

(b) Newton's second law for the astronaut is:

$$\Sigma(F_{\text{on A}})_y = n_{\text{C on A}} - (F_G)_A = m_A a_A \quad \Rightarrow \quad n_{\text{C on A}} = (F_G)_A + m_A a_A = m_A (g + a_A)$$

By Newton's third law, the astronaut's force on the chair is

$$n_{\text{A on C}} = n_{\text{C on A}} = m_A (g + a_A) = (80 \text{ kg})(9.8 \text{ m/s}^2 + 10 \text{ m/s}^2) = 1.6 \times 10^3 \text{ N}$$

Assess: This is a reasonable value because the astronaut's acceleration is greater than g.

7.11. Model: We treat the two objects of interest, the block (B) and steel cable (C), like particles. The motion of these objects is governed by the constant-acceleration kinematic equations. The horizontal component of the external force is 100 N.

Visualize:

Pictorial representation

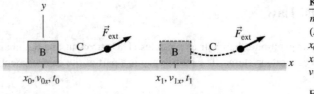

Known
$m_B = 20$ kg
$(F_{ext})_x = 100$ N
$x_0 = v_{0x} = t_0 = 0$
$x_1 = 2.0$ m
$v_{1x} = 4.0$ m/s

Find
m_C

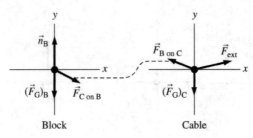

Block   Cable

Solve: Using $v_{1x}^2 = v_{0x}^2 + 2a_x(x_1 - x_0)$, we find

$$(4.0 \text{ m/s})^2 = 0 \text{ m}^2/\text{s}^2 + 2a_x(2.0 \text{ m}) \Rightarrow a_x = 4.0 \text{ m/s}^2$$

From the free-body diagram on the block:

$$\Sigma(F_{on B})_x = (F_{C on B})_x = m_B a_x \quad \Rightarrow \quad (F_{C on B})_x = (20 \text{ kg})(4.0 \text{ m/s}^2) = 80 \text{ N}$$

Also, according to Newton's third law $(F_{B on C})_x = (F_{C on B})_x = 80$ N. Applying Newton's second law to the cable gives

$$\Sigma(F_{on C})_x = (F_{ext})_x - (F_{B on C})_x = m_C a_x \quad \Rightarrow \quad 100 \text{ N} - 80 \text{ N} = m_C(4.0 \text{ m/s}^2) \quad \Rightarrow \quad m_C = 5.0 \text{ kg}$$

Section 7.4 Ropes and Pulleys

7.13. Model: The two ropes and the two blocks (A and B) will be treated as particles.
Visualize:

Free-body diagrams

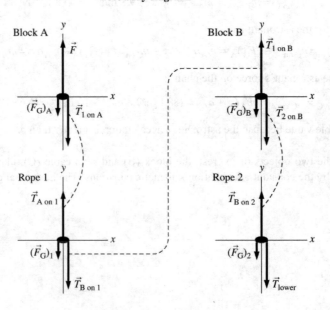

Solve: (a) The two blocks and two ropes form a combined system of total mass $M = 2.5$ kg. This combined system is accelerating upward at $a = 3.0$ m/s^2 under the influence of a force F and the gravitational force $-Mg\,\hat{\jmath}$. Newton's second law applied to the combined system gives

$$(F_{\text{net}})_y = F - Mg = Ma \quad \Rightarrow \quad F = M(a+g) = (2.5\text{ kg})(3.0\text{ m/s}^2 + 9.8\text{ m/s}^2) = 32 \text{ N}$$

(b) The ropes are *not* massless. We must consider both the blocks and the ropes as systems. The force F acts only on block A because it does not contact the other objects. We can proceed to apply the y-component of Newton's second law to each system, starting at the top. Each object accelerates upward at $a = 3.0$ m/s^2. For block A,

$$(F_{\text{net on A}})_y = F - m_A g - T_{1\text{ on A}} = m_A a \quad \Rightarrow \quad T_{1\text{ on A}} = F - m_A(a+g) = 19 \text{ N}$$

(c) Applying Newton's second law to rope 1 gives

$$(F_{\text{net on 1}})_y = T_{A\text{ on 1}} - m_1 g - T_{B\text{ on 1}} = m_1 a$$

where $\vec{T}_{A\text{ on 1}}$ and $\vec{T}_{1\text{ on A}}$ are an action/reaction pair. But, because the rope has mass, the two tension forces $\vec{T}_{A\text{ on 1}}$ and $\vec{T}_{B\text{ on 1}}$ are *not* the same. The tension at the lower end of rope 1, where it connects to B, is

$$T_{B\text{ on 1}} = T_{A\text{ on 1}} - m_1(a+g) = 16 \text{ N}$$

(d) We can continue to repeat this procedure, noting from Newton's third law that

$$T_{1\text{ on B}} = T_{B\text{ on 1}} \text{ and } T_{2\text{ on B}} = T_{B\text{ on 2}}$$

Newton's second law applied to block B is

$$(F_{\text{net on B}})_y = T_{1\text{ on B}} - m_B g - T_{2\text{ on B}} = m_B a \quad \Rightarrow \quad T_{2\text{ on B}} = T_{1\text{ on B}} - m_B(a+g) = 3.2 \text{ N}$$

7.17. Visualize:

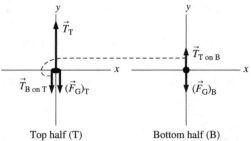

Top half (T)   Bottom half (B)

Solve: The rope is treated as two 1.0-kg interacting objects. At the midpoint of the rope, the rope has a tension $T_{B\text{ on T}} = T_{T\text{ on B}} \equiv T$. Apply Newton's first law to the bottom half of the rope to find T.

$$(F_{\text{net}})_y = 0 = T - (F_G)_B \quad \Rightarrow \quad T = m_B g = (1.0\text{ kg})(9.8\text{ m/s}^2) = 9.8 \text{ N}$$

Assess: 9.8 N is half the gravitational force on the whole rope. This is reasonable since the top half is holding up the bottom half of the rope against gravity.

7.21. Model: The block (B) and the steel cable (C), the two objects in the system, are modeled as particles and their motion is determined by the constant-acceleration kinematic equations.

Visualize:

Pictorial representation

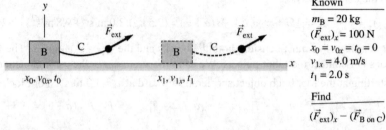

Known
$m_B = 20$ kg
$(\vec{F}_{ext})_x = 100$ N
$x_0 = v_{0x} = t_0 = 0$
$v_{1x} = 4.0$ m/s
$t_1 = 2.0$ s

Find
$(\vec{F}_{ext})_x - (\vec{F}_{B \text{ on } C})$

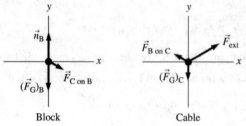

Block   Cable

Solve: Using $v_{1x} = v_{0x} + a_x(t_1 - t_0)$,

$$4.0 \text{ m/s} = 0 \text{ m/s} + a_x(2.0 \text{ s} - 0.0 \text{ s}) \quad \Rightarrow \quad a_x = 2.0 \text{ m/s}^2$$

Newton's second law along the x-direction for the block gives

$$\Sigma(F_{\text{on } B})_x = (F_{C \text{ on } B})_x = m_B a_x = (20 \text{ kg})(2.0 \text{ m/s}^2) = 40 \text{ N}$$

$(F_{ext})_x$ acts on the right end of the cable and $(F_{B \text{ on } C})_x$ acts on the left end. According to Newton's third law, $(F_{B \text{ on } C})_x = (F_{C \text{ on } B})_x = 40$ N. The difference in the horizontal component of the tension between the two ends of the cable is thus

$$(F_{ext})_x - (F_{B \text{ on } C})_x = 100 \text{ N} - 40 \text{ N} = 60 \text{ N}$$

7.23. Model: Sled A, sled B, and the dog (D) are treated like particles in the model. The rope is considered to be massless.
Visualize:

Pictorial representation

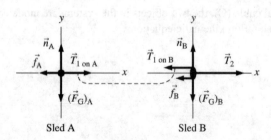

Known
$m_A = 100$ kg
$m_B = 80$ kg
$\mu_k = 0.10$
$T_1 = 150$ N

Find
T_2

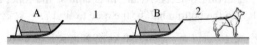

Sled A   Sled B

Solve: The acceleration constraint is $(a_A)_x = (a_B)_x = a_x$. Newton's second law applied to sled A gives

$$\Sigma(\vec{F}_{\text{on A}})_y = n_A - (F_G)_A = 0 \text{ N} \implies n_A = (F_G)_A = m_A g$$

$$\Sigma(\vec{F}_{\text{on A}})_x = T_{1 \text{ on A}} - f_A = m_A a_x$$

Using $f_A = \mu_k n_A$, the x-equation yields

$$T_{1 \text{ on A}} - \mu_k n_A = m_A a_x \implies 150 \text{ N} - (0.10)(100 \text{ kg})(9.8 \text{ m/s}^2) = (100 \text{ kg}) a_x \implies a_x = 0.52 \text{ m/s}^2$$

Newton's second law applied to sled B gives

$$\Sigma(\vec{F}_{\text{on B}})_y = n_B - (F_G)_B = 0 \text{ N} \implies n_B = (F_G)_B = m_B g$$

$$\Sigma(\vec{F}_{\text{on B}})_x = T_2 - T_{1 \text{ on B}} - f_B = m_B a_x$$

$T_{1 \text{ on B}}$ and $T_{1 \text{ on A}}$ act as if they are an action/reaction pair, so $T_{1 \text{ on B}} = 150 \text{ N}$. Using $f_B = \mu_k n_B = (0.10)(80 \text{ kg})$ $(9.8 \text{ m/s}^2) = 78.4 \text{ N}$, we find

$$T_2 - 150 \text{ N} - 78.4 \text{ N} = (80 \text{ kg})(0.52 \text{ m/s}^2) \implies T_2 = 270 \text{ N}$$

Thus the tension $T_2 = 2.7 \times 10^2 \text{ N}$.

7.27. Model: The starship and the shuttlecraft will be denoted as M and m, respectively, and both will be treated as particles. We will also use the constant-acceleration kinematic equations.
Visualize:

Pictorial representation

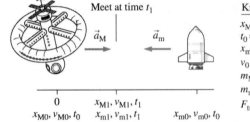
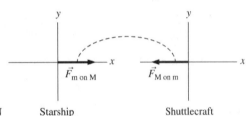

Solve: (a) The tractor beam is some kind of long-range force $\vec{F}_{\text{M on m}}$. Regardless of what kind of force it is, by Newton's third law there *must* be a reaction force $\vec{F}_{\text{m on M}}$ on the starship. As a result, both the shuttlecraft *and* the starship move toward each other (rather than the starship remaining at rest as it pulls the shuttlecraft in). However, the very different masses of the two crafts means that the distances they each move will also be very different. The pictorial representation shows that they meet at time t_1 when $x_{M1} = x_{m1}$. There's only one force on each craft, so Newton's second law is very simple. Furthermore, because the forces are an action/reaction pair,

$$F_{\text{M on m}} = F_{\text{m on M}} = F_{\text{tractor beam}} = 4.0 \times 10^4 \text{ N}$$

The accelerations of the two craft are

$$a_M = \frac{F_{\text{m on M}}}{M} = \frac{4.0 \times 10^4 \text{ N}}{2.0 \times 10^6 \text{ kg}} = 0.020 \text{ m/s}^2 \text{ and } a_m = \frac{\vec{F}_{\text{M on m}}}{m} = \frac{-4.0 \times 10^4 \text{ N}}{2.0 \times 10^4 \text{ kg}} = -2.0 \text{ m/s}^2$$

Acceleration a_m is negative because the force and acceleration vectors point in the negative x-direction. Now we have a constant-acceleration problem in kinematics. At a later time t_1 the positions of the crafts are

$$x_{M1} = x_{M0} + v_{M0}(t_1 - t_0) + \tfrac{1}{2} a_M(t_1 - t_0)^2 = \tfrac{1}{2} a_M t_1^2$$

$$x_{m1} = x_{m0} + v_{m0}(t_1 - t_0) + \tfrac{1}{2} a_m(t_1 - t_0)^2 = x_{m0} + \tfrac{1}{2} a_m t_1^2$$

The craft meet when $x_{M1} = x_{m1}$, so

$$\tfrac{1}{2} a_M t_1^2 = x_{m0} + \tfrac{1}{2} a_m t_1^2 \implies t_1 = \sqrt{\frac{2 x_{m0}}{a_M - a_m}} = \sqrt{\frac{2 x_{m0}}{a_M + |a_m|}} = \sqrt{\frac{2(10{,}000 \text{ m})}{2.02 \text{ m/s}^2}} = 99.5 \text{ s}$$

Knowing t_1, we can now find the starship's position as it meets the shuttlecraft:

$$x_{M1} = \tfrac{1}{2} a_M t_1^2 = 99 \text{ m}$$

The starship moves 99 m as it pulls in the shuttlecraft from 10 km away.

7.29. Model: The rock (R) and Bob (b) are the two objects in our system, and will be treated as particles. We will also use the constant-acceleration kinematic equations.
Visualize:

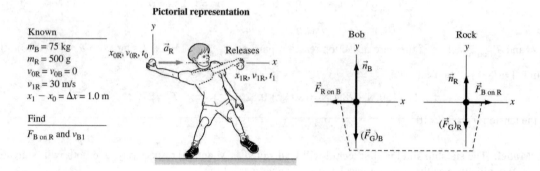

Solve: (a) Bob exerts a forward force $\vec{F}_{B \text{ on } R}$ on the rock to accelerate it forward. The rock's acceleration is calculated as follows:

$$v_{1R}^2 = v_{0R}^2 + 2a_{0R}\Delta x \quad \Rightarrow \quad a_R = \frac{v_{1R}^2}{2\Delta x} = \frac{(30 \text{ m/s})^2}{2(1.0 \text{ m})} = 450 \text{ m/s}^2$$

The force is calculated from Newton's second law:

$$F_{B \text{ on } R} = m_R a_R = (0.500 \text{ kg})(450 \text{ m/s}^2) = 225 \text{ N}$$

Bob exerts a force of 2.3×10^2 N on the rock.

(b) Because Bob pushes on the rock, the rock pushes back on Bob with a force $\vec{F}_{R \text{ on } B}$. Forces $\vec{F}_{R \text{ on } B}$ and $\vec{F}_{B \text{ on } R}$ are an action/reaction pair, so $F_{R \text{ on } B} = F_{B \text{ on } R} = 225$ N. The force causes Bob to accelerate backward with an acceleration of

$$a_B = \frac{(F_{\text{net on } B})_x}{m_B} = -\frac{F_{R \text{ on } B}}{m_B} = -\frac{225 \text{ N}}{75 \text{ kg}} = -3.0 \text{ m/s}^2$$

This is a rather large acceleration, but it lasts only until Bob releases the rock. We can determine the time interval by returning to the kinematics of the rock:

$$v_{1R} = v_{0R} + a_R \Delta t = a_R \Delta t \quad \Rightarrow \quad \Delta t = \frac{v_{1R}}{a_R} = 0.0667 \text{ s}$$

At the end of this interval, Bob's velocity is

$$v_{1B} = v_{0B} + a_B \Delta t = a_B \Delta t = -0.20 \text{ m/s}$$

Thus his recoil speed is 0.20 m/s.

7.31. Model: Assume package A and package B are particles. Use the model of kinetic friction and the constant-acceleration kinematic equations.

Visualize:

Pictorial representation

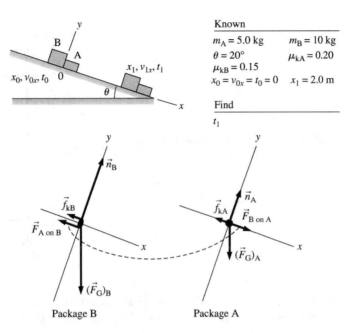

Known
$m_A = 5.0$ kg $m_B = 10$ kg
$\theta = 20°$ $\mu_{kA} = 0.20$
$\mu_{kB} = 0.15$
$x_0 = v_{0x} = t_0 = 0$ $x_1 = 2.0$ m

Find
t_1

Package B   Package A

Solve: Package B has a smaller coefficient of friction, so its acceleration down the ramp is greater than that of package A. It will therefore push against package A and, by Newton's third law, package A will push back on B. The acceleration constraint is $a_A = a_B \equiv a$.

Newton's second law applied to each package gives

$$\Sigma(F_{\text{on A}})_x = F_{\text{B on A}} + (F_G)_A \sin\theta - f_{kA} = m_A a$$

$$F_{\text{B on A}} + m_A g \sin\theta - \mu_{kA}(m_A g \cos\theta) = m_A a$$

$$\Sigma(F_{\text{on B}})_x = -F_{\text{A on B}} - f_{kB} + (F_G)_B \sin\theta = m_B a$$

$$-F_{\text{A on B}} - \mu_{kB}(m_B g \cos\theta) + m_B g \sin\theta = m_B a$$

where we have used $n_A = m_A \cos\theta g$ and $n_B = m_B \cos\theta g$. Adding the two force equations, and using $F_{\text{A on B}} = F_{\text{B on A}}$ because they are an action/reaction pair, we get

$$a = g\sin\theta - \frac{(\mu_{kA}m_A + \mu_{kB}m_B)(g\cos\theta)}{m_A + m_B} = \frac{[(020)(5.0 \text{ kg}) + (0.15)(10 \text{ kg})](9.8 \text{ m/s}^2)\cos(20°)}{5.0 \text{ kg} + 10 \text{ kg}} = 1.82 \text{ m/s}^2$$

Finally, using $x_1 = x_0 + v_0(t_1 - t_0) + \frac{1}{2}a(t_1 - t_0)^2$, we find

$$2.0 \text{ m} = 0 \text{ m} + 0 \text{ m} + \frac{1}{2}(1.82 \text{ m/s}^2)(t_1 - 0 \text{ s})^2 \quad \Rightarrow \quad t_1 = \sqrt{2(2.0 \text{ m})/(1.82 \text{ m/s}^2)} = 1.5 \text{ s}$$

7.35. Model: Blocks 1 and 2 make up the system of interest and will be treated as particles. Assume a massless rope and frictionless pulley.

Visualize:

Pictorial representation

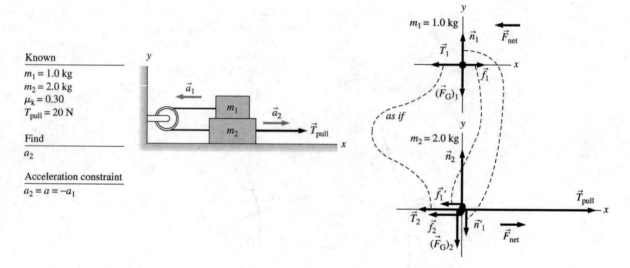

Known
$m_1 = 1.0$ kg
$m_2 = 2.0$ kg
$\mu_k = 0.30$
$T_{pull} = 20$ N

Find
a_2

Acceleration constraint
$a_2 = a = -a_1$

Solve: The blocks accelerate with the same magnitude but in opposite directions. Thus the acceleration constraint is $a_2 = a = -a_1$, where a will have a positive value. There are two real action/reaction pairs. The two tension forces will act as if they are action/reaction pairs because we are assuming a massless rope and a frictionless pulley. Make sure you understand why the friction forces point in the directions shown in the free-body diagrams, especially force $\vec{f}_1'$ exerted on block 2 by block 1. We have quite a few pieces of information to include. First, Newton's second law applied to blocks 1 and 2 gives

$$(\vec{F}_{\text{net on 1}})_x = f_1 - T_1 = \mu_k n_1 - T_1 = m_1 a_1 = -m_1 a$$

$$(F_{\text{net on 1}})_y = n_1 - m_1 g = 0 \text{ N} \implies n_1 = m_1 g$$

$$(F_{\text{net on 2}})_x = T_{\text{pull}} - f_1' - f_2 - T_2 = T_{\text{pull}} - f_1' - \mu_k n_2 - T_2 = m_2 a_2 = m_2 a$$

$$(F_{\text{net on 2}})_y = n_2 - n_1' - m_2 g = 0 \text{ N} \implies n_2 = n_1' + m_2 g$$

We've already used the kinetic friction model in both x-equations. Next, Newton's third law gives

$$n_1' = n_1 = m_1 g \quad f_1' = f_1 = \mu_k n_1 = \mu_k m_1 g \quad T_1 = T_2 = T$$

Knowing n_1', we can now use the y-equation of block 2 to find n_2. Substitute all these pieces into the two x-equations, and we end up with two equations with two unknowns:

$$\mu_k m_1 g - T = -m_1 a \quad T_{\text{pull}} - T - \mu_k m_1 g - \mu_k (m_1 + m_2) g = m_2 a$$

Subtract the first equation from the second to get

$$T_{\text{pull}} - \mu_k (3m_1 + m_2) g = (m_1 + m_2) a$$

$$a = \frac{T_{\text{pull}} - \mu_k (3m_1 + m_2) g}{m_1 + m_2} = \frac{20 \text{ N} - (0.30)[3(1.0 \text{ kg}) + 2.0 \text{ kg}](9.8 \text{ m/s}^2)}{1.0 \text{ kg} + 2.0 \text{ kg}} = 1.8 \text{ m/s}^2$$

7.37. Model: The sled (S) and the box (B) will be treated in the particle model, and the model of friction will be used. Refer to Table 6.1 for the required friction coefficients.

Visualize:

Pictorial representation

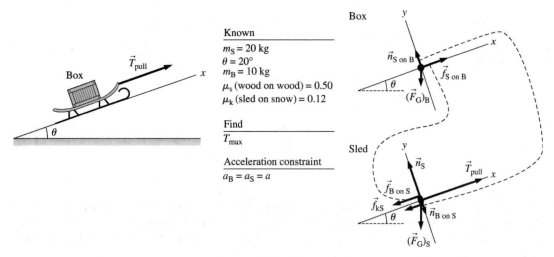

In the sled's free-body diagram n_S is the normal (contact) force on the sled due to the snow. Similarly f_{kS} is the force of kinetic friction on the sled due to snow.

Solve: Newton's second law on the box in the y-direction is

$$n_{S \text{ on } B} - (F_G)_B \cos(20°) = 0 \text{ N} \quad \Rightarrow \quad n_{S \text{ on } B} = (10 \text{ kg})(9.8 \text{ m/s}^2)\cos(20°) = 92.1 \text{ N}$$

The static friction force $\vec{f}_{S \text{ on } B}$ accelerates the box. The maximum acceleration occurs when static friction reaches its maximum possible value.

$$(f_s)_{\text{max}} = \mu_s n_{S \text{ on } B} = (0.50)(92.1 \text{ N}) = 46.1 \text{ N}$$

Newton's second law along the x-direction thus gives the maximum acceleration

$$f_{S \text{ on } B} - (F_G)_B \sin(20°) = m_B a \quad \Rightarrow \quad 46.1 \text{ N} - (10 \text{ kg})(9.8 \text{ m/s}^2)\sin(20°) = (10 \text{ kg})a \quad \Rightarrow \quad a = 1.25 \text{ m/s}^2$$

Newton's second law for the sled along the y-direction is

$$n_S - n_{B \text{ on } S} - (F_G)_S \cos(20°) = 0 \text{ N}$$

$$n_S = n_{B \text{ on } S} + m_S g \cos(20°) = (92.1 \text{ N}) + (20 \text{ kg})(9.8 \text{ m/s}^2)\cos(20°) = 276.3 \text{ N}$$

Therefore, the force of friction on the sled by the snow is

$$f_{kS} = (\mu_k) n_S = (0.06)(276.3 \text{ N}) = 16.6 \text{ N}$$

Newton's second law along the x-direction is

$$T_{\text{pull}} - w_S \sin(20°) - f_{kS} - f_{B \text{ on } S} = m_S a$$

The friction force $f_{B \text{ on } S} = f_{S \text{ on } B}$ because these are an action/reaction pair. We're using the maximum acceleration, so the maximum tension is

$$T_{\text{max}} - (20 \text{ kg})(9.8 \text{ m/s}^2)\sin(20°) - 16.6 \text{ N} - 46.1 \text{ N} = (20 \text{ kg})(1.25 \text{ m/s}^2) = 160 \text{ N}$$

7.41. Model: Assume the particle model for the two blocks, and the model of kinetic and static friction.

Visualize:

Pictorial representation

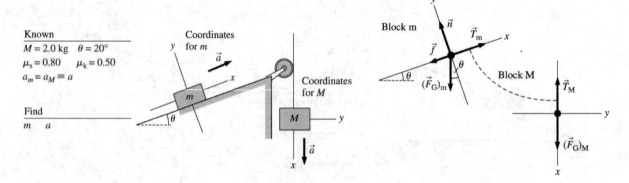

Solve: **(a)** If the mass m is too small, the hanging 2.0 kg mass will pull it up the slope. We want to find the smallest mass that will stick as a result of friction. The smallest mass will be the one for which the force of static friction is at its maximum possible value: $f_s = (f_s)_{max} = \mu_s n$. As long as the mass m is stuck, both blocks are at rest with $\vec{F}_{net} = 0$ N. In this situation, Newton's second law for the hanging mass M gives

$$(F_{net})_x = -T_M + Mg = 0 \text{ N} \quad \Rightarrow \quad T_M = Mg = (2.0 \text{ kg})(9.8 \text{ m/s}^2) = 19.6 \text{ N}$$

For the smaller mass m,

$$(F_{net})_x = T_m - f_s - mg\sin\theta = 0 \text{ N} \quad (F_{net})_y = n - mg\cos\theta \quad \Rightarrow \quad n = mg\cos\theta$$

For a massless string and frictionless pulley, forces $\vec{T}_m$ and $\vec{T}_M$ act as if they are an action/reaction pair. Thus $T_m = T_M$. Mass m is a minimum when $f_s = (f_s)_{max} = \mu_s n = \mu_s mg\cos\theta$. Substituting these expressions into the x-equation for m gives

$$T_M - \mu_s mg\cos\theta - mg\sin\theta = 0 \text{ N}$$

$$m = \frac{T_M}{(\mu_s\cos\theta + \sin\theta)g} = \frac{19.6 \text{ N}}{[(0.80)\cos(20°) + \sin(20°)](9.8 \text{ m/s}^2)} = 1.83 \text{ kg}$$

or 1.8 kg to two significant figures.

(b) Because $\mu_k < \mu_s$ the 1.8 kg block will begin to slide up the ramp and the 2.0 kg mass will begin to fall if the block is nudged ever so slightly. In this case, the net force and the acceleration are *not* zero. Notice how, in the pictorial representation, we chose different coordinate systems for the two masses. The magnitudes of the accelerations are the same because the blocks are tied together. Thus, the acceleration constraint is $a_m = a_M \equiv a$, where a will have a positive value. Newton's second law for block M gives

$$(F_{net})_x = -T + Mg = Ma_M = Ma$$

For block m we have

$$(F_{net})_x = T - f_k - mg\sin\theta = T - \mu_k mg\cos\theta - mg\sin\theta = ma_m = ma$$

In writing these equations, we used Newton's third law to obtain $T_m = T_M = T$. Also, notice that the x-equation and the friction model for block m don't change, except for μ_s becoming μ_k, so we already know the expression for f_k from part (a). Notice that the tension in the string is *not* the gravitational force Mg. We have two equations with the two unknowns T and a:

$$Mg - T = Ma \quad T - (\mu_k\cos\theta + \sin\theta)mg = ma$$

Adding the two equations to eliminate T gives

$$Mg - (\mu_k \cos\theta + \sin\theta)mg = Ma + ma$$

$$a = g\frac{M - (\mu_k \cos\theta + \sin\theta)m}{M + m}$$

$$= (9.8 \text{ m/s}^2)\frac{2.0 \text{ kg} - [(0.50)\cos(20°) + \sin(20°)](1.83 \text{ kg})}{2.0 \text{ kg} + 1.83 \text{ kg}} 1.3 \text{ m/s}^2$$

7.45. Model: Assume the cable mass is negligible compared to the car mass and that the pulley is frictionless. Use the particle model for the two cars.

Visualize: Please refer to Figure P7.45.

Pictorial representation

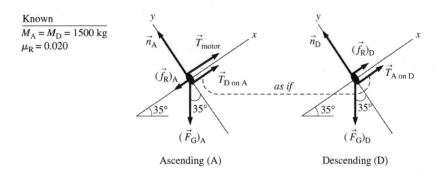

Known
$M_A = M_D = 1500 \text{ kg}$
$\mu_R = 0.020$

Ascending (A)   Descending (D)

Solve: (a) The cars are moving at constant speed, so they are in dynamic equilibrium. Consider the descending car D. We can find the rolling friction force on car D, and then find the cable tension by applying Newton's first law. In the y-direction for car D,

$$(F_{net})_y = 0 = n_D - (F_G)_D \cos(35°)$$

$$n_D = m_D g \cos(35°)$$

So the rolling friction force on car D is

$$(f_R)_D = \mu_R n_D = \mu_R m_D g \cos(35°)$$

Applying Newton's first law to car D in the x-direction gives

$$(F_{net})_x = T_{A \text{ on } D} + (f_R)_D - (F_G)_D \sin(35°) = 0$$

Thus,

$$T_{A \text{ on } D} = m_D g \sin(35°) - \mu_R m_D g \cos(35°)$$

$$= (1500 \text{ kg})(9.80 \text{ m/s}^2)[\sin(35°) - (0.020)\cos(35°)]$$

$$= 8.2 \times 10^3 \text{ N}$$

(b) Similarly, we find that for car A, $(f_R)_A = \mu_R m_A g \cos(35°)$. In the x-direction for car A,

$$(F_{net})_x = T_{motor} + T_{D \text{ on } A} - (f_R)_A - (F_G)_A \sin(35°) = 0$$

$$T_{motor} = m_A g \sin(35°) + \mu_R m_A g \cos(35°) - m_D g \sin(35°) + \mu_R m_D g \cos(35°)$$

Here, we have used $T_{A \text{ on } D} = T_{D \text{ on } A}$. If we also use $m_A = m_D$, then

$$T_{motor} = 2\mu_R m_A g \cos(35°) = 4.8 \times 10^2 \text{ N}.$$

Assess: Careful examination of the free-body diagrams for cars D and A yields the observation that $T_{motor} = 2(F_R)_A$ in order for the cars to be in dynamic equilibrium. It is a tribute to the design that the motor must only provide such a small force compared to the tension in the cable connecting the two cars.

7.47. Model: Model Jorge as a particle and use the friction model.
Visualize:

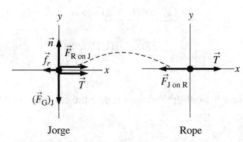

Jorge   Rope

Solve: If the Jorge pulls on the rope with force F, Newton's third law requires the rope to pull up on Jorge with force F. This is just the tension in the rope (i.e., $F = T$). With our model of the rope and pulley, the same tension pulls at Jorge's waist where the rope is tied. Applying Newton's second law to Jorge in the y-direction gives

$$\sum(F)_y = n - F_G = 0 \implies n = F_G = mg$$

From the friction model, we have $f_r = \mu_r n = \mu_r mg$. Using this in the equation below, which is derived by using Newton's second law applied in the x-direction, gives

$$\sum(F)_x = F + T - f_r = ma$$

$$2F - \mu_r mg_r = ma$$

$$a = \frac{2F}{m} - \mu_r g$$

7.49. Model: Use the particle model for the wedge and the block.
Visualize:

Pictorial representation

Known
m_1 m_2 θ

Find
F for no slip

Block   Wedge

The block will not slip relative to the wedge if they both have the same horizontal acceleration a. Note that $n_{1 \text{ on } 2}$ and $n_{2 \text{ on } 1}$ form a third law pair, so $n_{1 \text{ on } 2} = n_{2 \text{ on } 1}$.
Solve: Newton's second law applied to block m_2 in the y-direction gives

$$\sum(F_{\text{on } 2})_y = n_{1 \text{ on } 2} \cos\theta - (F_G)_2 = 0 \text{ N} \implies n_{1 \text{ on } 2} = \frac{m_2 g}{\cos\theta}$$

Combining this equation with the x-component of Newton's second law yields:

$$\sum(F_{\text{on } 2})_x = n_{1 \text{ on } 2} \sin\theta = m_2 a \implies a = \frac{n_{1 \text{ on } 2} \sin\theta}{m_2} = g \tan\theta$$

Newton's second law applied to the wedge gives

$$\sum(F_{\text{on } 1})_x = F - n_{2 \text{ on } 1} \sin\theta = m_1 a$$

$$F = m_1 a + n_{2 \text{ on } 1} \sin\theta = m_1 a + m_2 a = (m_1 + m_2)a = (m_1 + m_2)g \tan\theta$$

DYNAMICS II: MOTION IN A PLANE

Exercises and Problems

Section 8.1 Dynamics in Two Dimensions

8.1. Model: The model rocket and the target will be treated as particles. The kinematics equations in two dimensions apply.

Visualize:

Pictorial representation

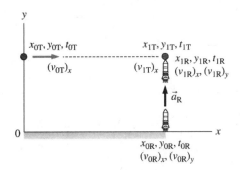

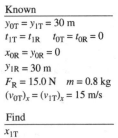

Known
$y_{0T} = y_{1T} = 30$ m
$t_{1T} = t_{1R}$ $t_{0T} = t_{0R} = 0$
$x_{0R} = y_{0R} = 0$
$y_{1R} = 30$ m
$F_R = 15.0$ N $m = 0.8$ kg
$(v_{0T})_x = (v_{1T})_x = 15$ m/s

Find
x_{1T}

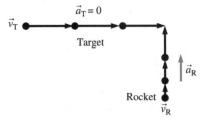

Solve: For the rocket, Newton's second law along the y-direction is

$$(F_{net})_y = F_R - mg = ma_R$$

$$\Rightarrow a_R = \frac{1}{m}(F_R - mg) = \frac{1}{0.8 \text{ kg}}[15 \text{ N} - (0.8 \text{ kg})(9.8 \text{ m/s}^2)] = 8.95 \text{ m/s}^2$$

Using the kinematic equation $y_{1R} = y_{0R} + (v_{0R})_y(t_{1R} - t_{0R}) + \frac{1}{2}a_R(t_{1R} - t_{0R})^2$,

$$30 \text{ m} = 0 \text{ m} + 0 \text{ m} + \frac{1}{2}(8.95 \text{ m/s}^2)(t_{1R} - 0 \text{ s})^2 \Rightarrow t_{1R} = 2.589 \text{ s}$$

For the target (noting $t_{1T} = t_{1R}$),

$$x_{1T} = x_{0T} + (v_{0T})_x(t_{1T} - t_{0T}) + \frac{1}{2}a_T(t_{1T} - t_{0T})^2 = 0 \text{ m} + (15 \text{ m/s})(2.589 \text{ s} - 0 \text{ s}) + 0 \text{ m} = 39 \text{ m}$$

You should launch when the target is 39 m away.

Assess: The rocket is to be fired when the target is at x_{0T}. For a net acceleration of approximately 9 m/s² in the vertical direction and a time of 2.6 s to cover a vertical distance of 30 m, a horizontal distance of 39 m is reasonable.

8.3. Model: The asteroid and the giant rocket will be treated as particles undergoing motion according to the constant-acceleration equations of kinematics.
Visualize:

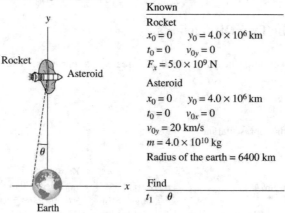

Pictorial representation

Rocket

Asteroid

Earth

Known
Rocket
$x_0 = 0$ $y_0 = 4.0 \times 10^6$ km
$t_0 = 0$ $v_{0y} = 0$
$F_x = 5.0 \times 10^9$ N

Asteroid
$x_0 = 0$ $y_0 = 4.0 \times 10^6$ km
$t_0 = 0$ $v_{0x} = 0$
$v_{0y} = 20$ km/s
$m = 4.0 \times 10^{10}$ kg
Radius of the earth = 6400 km

Find
t_1 θ

Solve: **(a)** The time it will take the asteroid to reach the earth is

$$\frac{\text{displacement}}{\text{velocity}} = \frac{4.0 \times 10^6 \text{ km}}{20 \text{ km/s}} = 2.0 \times 10^5 \text{ s} = 56 \text{ h}$$

(b) The angle of a line that just misses the earth is

$$\tan\theta = \frac{R}{y_0} \Rightarrow \theta = \tan^{-1}\left(\frac{R}{y_0}\right) = \tan^{-1}\left(\frac{6400 \text{ km}}{4.0 \times 10^6 \text{ km}}\right) = 0.092°$$

(c) When the rocket is fired, the horizontal acceleration of the asteroid is

$$a_x = \frac{5.0 \times 10^9 \text{ N}}{4.0 \times 10^{10} \text{ kg}} = 0.125 \text{ m/s}^2$$

(Note that the mass of the rocket is much smaller than the mass of the asteroid and can therefore be ignored completely.) The velocity of the asteroid after the rocket has been fired for 300 s is

$$v_x = v_{0x} + a_x(t - t_0) = 0 \text{ m/s} + (0.125 \text{ m/s}^2)(300 \text{ s} - 0 \text{ s}) = 37.5 \text{ m/s}$$

After 300 s, the vertical velocity is $v_y = 2 \times 10^4$ m/s and the horizontal velocity is $v_x = 37.5$ m/s. The deflection due to this horizontal velocity is

$$\tan\theta = \frac{v_x}{v_y} \Rightarrow \theta = \tan^{-1}\left(\frac{37.5 \text{ m/s}}{2 \times 10^4 \text{ m/s}}\right) = 0.107°$$

That is, the earth is saved.

Section 8.2 Uniform Circular Motion

8.7. Solve: Newton's second law is $F_r = ma_r = mr\omega^2$. Substituting into this equation yields:

$$\omega = \sqrt{\frac{F_r}{mr}} = \sqrt{\frac{8.2 \times 10^{-8} \text{ N}}{(9.1 \times 10^{-31} \text{ kg})(5.3 \times 10^{-11} \text{ m})}}$$

$$= 4.37 \times 10^{16} \text{ rad/s} = 4.37 \times 10^{16} \frac{\text{rad}}{\text{s}} \times \frac{1 \text{ rev}}{2\pi \text{ rad}} = 6.6 \times 10^{15} \text{ rev/s}$$

Assess: This is a very high number of revolutions per second.

8.9. Model: The motion of the moon around the earth will be treated through the particle model. The circular motion is uniform.
Visualize:

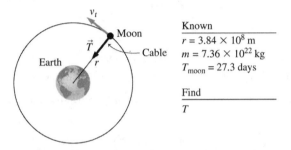

Pictorial representation

Known
$r = 3.84 \times 10^8$ m
$m = 7.36 \times 10^{22}$ kg
$T_{\text{moon}} = 27.3$ days

Find
T

Solve: The tension in the cable provides the centripetal acceleration. Newton's second law is

$$\Sigma F_r = T = mr\omega^2 = mr\left(\frac{2\pi}{T_{\text{moon}}}\right)^2$$

$$= (7.36 \times 10^{22} \text{ kg})(3.84 \times 10^8 \text{ m})\left[\frac{2\pi}{27.3 \text{ days}} \times \frac{1 \text{ day}}{24 \text{ h}} \times \frac{1 \text{ h}}{3600 \text{ s}}\right]^2 = 2.01 \times 10^{20} \text{ N}$$

Assess: This is a tremendous tension, but clearly understandable in view of the moon's large mass and the large radius of circular motion around the earth. This is the same answer we'll get later with Newton's law of universal gravitation.

Section 8.4 Fictitious Forces

8.13. Model: Use the particle model for the car which is undergoing circular motion.
Visualize:

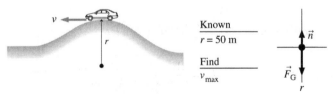

Pictorial representation

Known
$r = 50$ m

Find
v_{max}

Solve: The car is in circular motion with the center of the circle below the car. Newton's second law at the top of the hill is

$$\Sigma F_r = (F_G)_r - n_r = mg - n = ma_r = \frac{mv^2}{r} \Rightarrow v^2 = r\left(g - \frac{n}{m}\right)$$

Maximum speed is reached when $n = 0$ and the car is beginning to lose contact with the road.

$$v_{\text{max}} = \sqrt{rg} = \sqrt{(50 \text{ m})(9.8 \text{ m/s}^2)} = 22 \text{ m/s}$$

Assess: A speed of 22 m/s is equivalent to 49 mph, which seems like a reasonable value.

8.15. Model: Model the roller coaster car as a particle at the top of a circular loop-the-loop undergoing uniform circular motion.
Visualize:

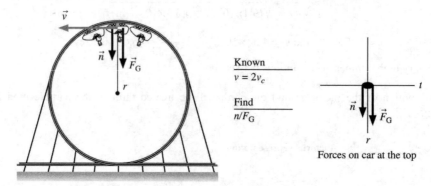

Pictorial representation

Known
$v = 2v_c$

Find
n/F_G

Forces on car at the top

Notice that the r-axis points downward, toward the center of the circle.
Solve: The critical speed occurs when $\vec{n}$ goes to zero and $\vec{F}_G$ provides all the centripetal force pulling the car in the vertical circle. At the critical speed $mg = mv_c^2/r$, therefore $v_c = \sqrt{rg}$. Since the car's speed is twice the critical speed, $v_t = 2v_c$ and the centripetal force is

$$\Sigma F_r = n + F_G = \frac{mv^2}{r} = \frac{m(4v_c^2)}{r} = \frac{m(4rg)}{r} = 4mg$$

Thus the normal force is $n = 3\,mg$. Consequently, $n/F_G = 3$.

Section 8.5 Nonuniform Circular Motion

8.19. Model: Use the particle model for the car, which is undergoing nonuniform circular motion.
Visualize:

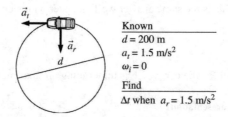

Known
$d = 200$ m
$a_t = 1.5$ m/s^2
$\omega_i = 0$

Find
Δt when $a_r = 1.5$ m/s^2

Solve: The car is in circular motion with radius $r = \dfrac{d}{2} = 100$ m. We require

$$a_r = \omega^2 r = 1.5 \text{ m/s}^2 \Rightarrow \omega = \sqrt{\frac{1.5 \text{ m/s}^2}{r}} = \sqrt{\frac{1.5 \text{ m/s}^2}{100 \text{ m}}} = 0.122 \text{ s}^{-1}$$

The definition of the angular velocity can be used to determine the time Δt using the angular acceleration $\alpha = \dfrac{a_t}{r} = \dfrac{1.5 \text{ m/s}^2}{100 \text{ m}} = 1.5 \times 10^{-2} \text{ s}^{-2}$.

$$\omega = \omega_i + \alpha \Delta t$$

$$\Rightarrow \Delta t = \frac{\omega - \omega_i}{\alpha} = \frac{0.122 \text{ s}^{-1} - 0 \text{ s}^{-1}}{0.015 \text{ s}^{-2}} = 8.2 \text{ s}$$

8.21. Model: The object is treated as a particle in the model of kinetic friction with its motion governed by constant-acceleration kinematics.
Visualize:

Pictorial representation

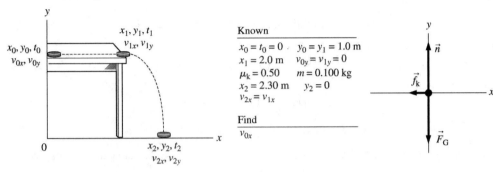

Known

$x_0 = t_0 = 0$ $y_0 = y_1 = 1.0$ m
$x_1 = 2.0$ m $v_{0y} = v_{1y} = 0$
$\mu_k = 0.50$ $m = 0.100$ kg
$x_2 = 2.30$ m $y_2 = 0$
$v_{2x} = v_{1x}$

Find

v_{0x}

Solve: The velocity v_{1x} as the object sails off the edge is related to the initial velocity v_{0x} by $v_{1x}^2 = v_{0x}^2 + 2a_x(x_1 - x_0)$. Using Newton's second law to determine a_x while sliding gives

$$\sum F_x = -f_k = ma_x \Rightarrow \sum F_y = n - mg = 0 \text{ N} \Rightarrow n = mg$$

Using this result and the model of kinetic friction ($f_k = \mu_k n$), the x-component equation can be written as $-\mu_k mg = ma_x$. This implies

$$a_x = -\mu_k g = -(0.50)(9.8 \text{ m/s}^2) = -4.9 \text{ m/s}^2$$

Kinematic equations for the object's free fall can be used to determine v_{1x}:

$$y_2 = y_1 + v_{1y}(t_2 - t_1) + \tfrac{1}{2}(-g)(t_2 - t_1)^2 \Rightarrow 0 \text{ m} = 1.0 \text{ m} + 0 \text{ m} - \frac{g}{2}(t_2 - t_1)^2 \Rightarrow (t_2 - t_1) = 0.4518 \text{ s}$$

$$x_2 = x_1 + v_{1x}(t_2 - t_1) = 2.30 \text{ m} = 2.0 \text{ m} + v_{1x}(0.4518 \text{ s}) \Rightarrow v_{1x} = 0.664 \text{ m/s}$$

Having determined v_{1x} and a_x, we can go back to the velocity equation $v_{1x}^2 = v_{0x}^2 + 2a_x(x_1 - x_0)$:

$$(0.664 \text{ m/s})^2 = v_{0x}^2 + 2(-4.9 \text{ m/s}^2)(2.0 \text{ m}) \Rightarrow v_{0x} = 4.5 \text{ m/s}$$

Assess: $v_{0x} = 4.5$ m/s is about 10 mph and is a reasonable speed.

8.27. Model: The model rocket is treated as a particle and its motion is determined by constant-acceleration kinematic equations.
Visualize:

Pictorial representation

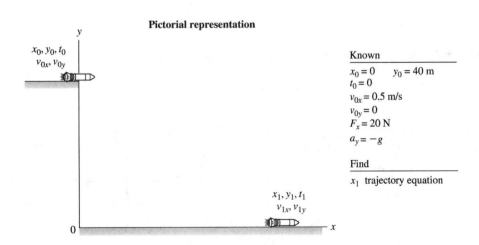

Known

$x_0 = 0$ $y_0 = 40$ m
$t_0 = 0$
$v_{0x} = 0.5$ m/s
$v_{0y} = 0$
$F_x = 20$ N
$a_y = -g$

Find

x_1 trajectory equation

Solve: As the rocket is accidentally bumped $v_{0x} = 0.5$ m/s and $v_{0y} = 0$ m/s. On the other hand, when the engine is fired

$$F_x = ma_x \Rightarrow a_x = \frac{F_x}{m} = \frac{20 \text{ N}}{0.500 \text{ kg}} = 40 \text{ m/s}^2$$

(a) Using $y_1 = y_0 + v_{0y}(t_1 - t_0) + \frac{1}{2}a_y(t_1 - t_0)^2$,

$$0 \text{ m} = 40 \text{ m} + 0 \text{ m} + \frac{1}{2}(-9.8 \text{ m/s}^2)t_1^2 \Rightarrow t_1 = 2.857 \text{ s}$$

The distance from the base of the wall is

$$x_1 = x_0 + v_{0x}(t_1 - t_0) + \frac{1}{2}a_x(t_1 - t_0)^2 = 0 \text{ m} + (0.5 \text{ m/s})(2.857\text{s}) + \frac{1}{2}(40 \text{ m/s}^2)(2.857 \text{ s})^2 = 165 \text{ m}$$

(b) The x- and y-equations are

$$y = y_0 + v_{0y}(t - t_0) + \frac{1}{2}a_y(t - t_0)^2 = 40 - 4.9t^2$$

$$x = x_0 + v_{0x}(t - t_0) + \frac{1}{2}a_x(t - t_0)^2 = 0.5t + 20t^2$$

Except for a brief interval near $t = 0$, $20t^2 \gg 0.5t$. Thus $x \approx 20t^2$, or $t^2 = x/20$. Substituting this into the y-equation gives

$$y = 40 - 0.245x$$

This is the equation of a straight line, so the rocket follows a linear trajectory to the ground.

8.29. Model: Assume the particle model for the satellite in circular motion.
Visualize:

Pictorial representation

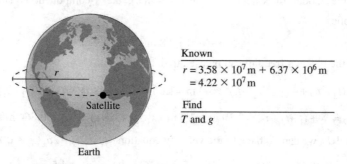

Known

$r = 3.58 \times 10^7 \text{ m} + 6.37 \times 10^6 \text{ m}$
$\quad = 4.22 \times 10^7 \text{ m}$

Find

T and g

To be in a geosynchronous orbit means rotating at the same rate as the earth, which is 24 hours for one complete rotation. Because the altitude of the satellite is 3.58×10^7 m, $r = 3.58 \times 10^7$ m, $r_e = 3.58 \times 10^7$ m $+ 6.37 \times 10^6$ m $= 4.22 \times 10^7$ m.
Solve: (a) The period (T) of the satellite is 24.0 hours.
(b) The acceleration due to gravity is

$$g = a_r = r\omega^2 = r\left(\frac{2\pi}{T}\right)^2 = (4.22 \times 10^7 \text{m})\left(\frac{2\pi}{24.0 \text{ hr}} \times \frac{1 \text{ hr}}{3600 \text{ s}}\right)^2 = 0.223 \text{ m/s}^2$$

(c) There is no normal force on a satellite, so the weight is zero. It is in free fall.

8.31. Model: Model the ball as a particle which is in a vertical circular motion.
Visualize:

Pictorial representation

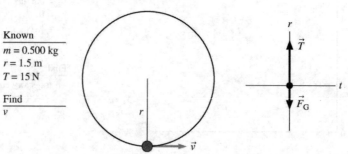

Known
$m = 0.500$ kg
$r = 1.5$ m
$T = 15$ N

Find
v

Solve: At the bottom of the circle,

$$\sum F_r = T - F_G = \frac{mv^2}{r} \Rightarrow (15 \text{ N}) - (0.500 \text{ kg})(9.8 \text{ m/s}^2) = \frac{(0.500 \text{ kg})v^2}{(1.5 \text{ m})} \Rightarrow v = 5.5 \text{ m/s}$$

8.35. Model: Use the particle model for the car, which is in uniform circular motion.
Visualize:

Pictorial representation

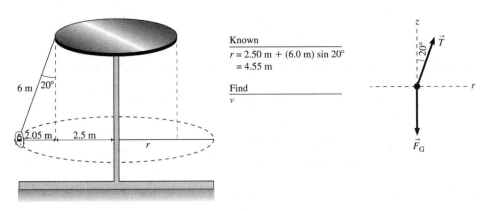

Known
$r = 2.50 \text{ m} + (6.0 \text{ m}) \sin 20°$
$= 4.55 \text{ m}$

Find
v

Solve: Newton's second law is

$$\sum F_r = T \sin 20° = ma_r = \frac{mv^2}{r} \quad \sum F_z = T \cos 20° - F_G = 0 \text{ N}$$

These equations can be written as

$$T \sin 20° = \frac{mv^2}{r} \quad T \cos 20° = mg$$

Dividing these two equations gives

$$\tan 20° = v^2/rg \Rightarrow v = \sqrt{rg \tan 20°} = \sqrt{(4.55 \text{ m})(9.8 \text{ m/s}^2) \tan 20°} = 4.03 \text{ m/s} \approx 4 \text{ m/s}$$

8.37. Model: Assume the particle model for a sphere in circular motion at constant speed.
Visualize:

Pictorial representation

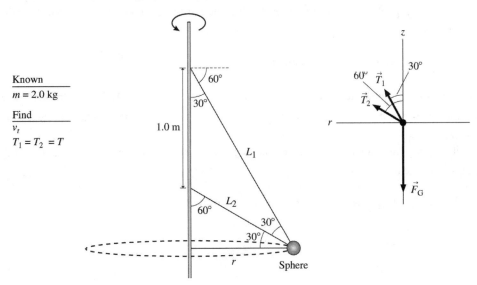

Known
$m = 2.0 \text{ kg}$

Find
v_t
$T_1 = T_2 = T$

Solve: **(a)** Newton's second law along the r and z axes is:

$$\Sigma F_r = T_1 \sin 30° + T_2 \sin 60° = \frac{mv_t^2}{r} \quad \Sigma F_z = T_1 \cos 30° + T_2 \cos 60° - F_G = 0 \text{ N}$$

Since we want $T_1 = T_2 = T$, these two equations become

$$T(\sin 30° + \sin 60°) = \frac{mv_t^2}{r} \quad T(\cos 30° + \cos 60°) = mg$$

Since $\sin 30° + \sin 60° = \cos 30° + \cos 60°$,

$$mg = \frac{mv_t^2}{r} \Rightarrow v_t = \sqrt{rg}$$

The triangle with sides L_1, L_2, and 1.0 m is isosceles, so $L_2 = 1.0$ m and $r = L_2 \cos 30°$. Thus

$$\sqrt{L_2 \cos 30° g} = \sqrt{(1.0\,\text{m})\cos 30° g} = \sqrt{(0.866\,\text{m})(9.8\,\text{m/s}^2)} = 2.9 \text{ m/s}$$

(b) The tension is

$$T = \frac{mg}{\cos 30° + \cos 60°} = \frac{(2.0 \text{ kg})(9.8 \text{ m/s}^2)}{0.866 + 0.5} = 14.3 \text{ N} \approx 14 \text{ N}$$

8.39. Model: Use the particle model for the marble in uniform circular motion.
Visualize:

Pictorial representation

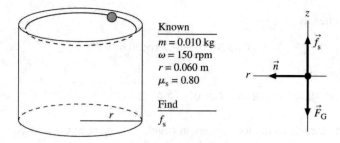

Known
$m = 0.010$ kg
$\omega = 150$ rpm
$r = 0.060$ m
$\mu_s = 0.80$

Find
f_s

Solve: The marble will roll in a horizontal circle if the static friction force is sufficient to support the gravitational on it: $f_s = F_G$. If $mg > (f_s)_{max}$ then static friction is not sufficient and the marble will slip down the side as it rolls around the circumference. The r-equation of Newton's second law is

$$\Sigma F_r = n = mr\omega^2 = (0.010 \text{ kg})(0.060 \text{ m})\left(150 \text{ rpm} \times \frac{2\pi \text{ rad}}{1 \text{ rev}} \times \frac{1 \text{ min}}{60 \text{ s}}\right)^2 = 0.148 \text{ N}$$

Thus the maximum possible static friction is $(f_s)_{max} = \mu_s n = (0.80)(0.148 \text{ N}) = 0.118 \text{ N}$. The friction force needed to support a 10 g marble is $f_s = mg = 0.098$ N. We see that $f_s < (f_s)_{max}$, therefore friction is sufficient and the marble spins in a horizontal circle.
Assess: In reality, rolling friction will cause the marble to gradually slow down until $(f_s)_{max} < mg$. At that point, it will begin to slip down the inside wall.

8.43. Model: Model a passenger as a particle rotating in a vertical circle.

Visualize:

Pictorial representation

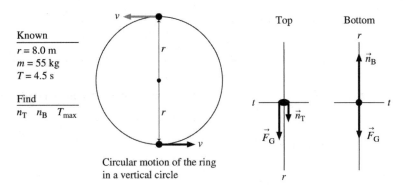

Known

$r = 8.0$ m
$m = 55$ kg
$T = 4.5$ s

Find

n_T n_B T_{max}

Circular motion of the ring
in a vertical circle

Top   Bottom

Solve: **(a)** Newton's second law at the top is

$$\sum F_r = n_T + F_G = ma_r = \frac{mv^2}{r} \Rightarrow n_T + mg = \frac{mv^2}{r}$$

The speed is

$$v = \frac{2\pi r}{T} = \frac{2\pi(8.0 \text{ m})}{4.5 \text{ s}} = 11.17 \text{ m/s}$$

$$\Rightarrow n_T = m\left(\frac{v^2}{r} - g\right) = (55 \text{ kg})\left[\frac{(11.17 \text{ m/s})^2}{8.0 \text{ m}} - 9.8 \text{ m/s}^2\right] = 319 \text{ N}$$

That is, the ring pushes on the passenger with a force of 3.2×10^2 N at the top of the ride. Newton's second law at the bottom:

$$\sum F_r = n_B - F_G = ma_r = \frac{mv^2}{r} \Rightarrow n_B = \frac{mv^2}{r} + mg = m\left(\frac{v^2}{r} + g\right)$$

$$= (55 \text{ kg})\left[\frac{(11.17 \text{ m/s})^2}{8.0 \text{ m}} + 9.8 \text{ m/s}^2\right] = 1397 \text{ N}$$

Thus the force with which the ring pushes on the rider when she is at the bottom of the ring is 1.4 kN.
(b) To just stay on at the top, $n_T = 0$ N in the r-equation at the top in part (a). Thus,

$$mg = \frac{mv^2}{r} = mr\omega^2 = mr\left(\frac{2\pi}{T_{max}}\right)^2 \Rightarrow T_{max} = 2\pi\sqrt{\frac{r}{g}} = 2\pi\sqrt{\frac{8.0 \text{ m}}{9.8 \text{ m/s}^2}} = 5.7 \text{ s}$$

8.47. Model: Model the ball as a particle in uniform circular motion. Rolling friction is ignored.
Visualize:

Pictorial representation

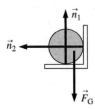

Solve: The track exerts both an upward normal force and an inward normal force. From Newton's second law,

$$n_1 = mg = (0.030 \text{ kg})(9.8 \text{ m/s}^2) = 0.294 \text{ N, up}$$

$$n_2 = mr\omega^2 = (0.030 \text{ kg})(0.20 \text{ m})\left[\frac{60 \text{ rev}}{\text{min}} \times \frac{2\pi \text{ rad}}{1 \text{ rev}} \times \frac{1 \text{ min}}{60 \text{ s}}\right]^2 = 0.2369 \text{ N, in}$$

$$F_{\text{net}} = \sqrt{n_1^2 + n_2^2} = \sqrt{(0.294 \text{ N})^2 + (0.2369 \text{ N})^2} = 0.38 \text{ N}$$

8.51. Model: Use the particle model for a ball in motion in a vertical circle and then as a projectile.
Visualize:

Pictorial representation

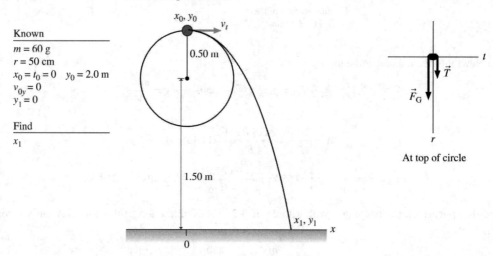

Known
$m = 60$ g
$r = 50$ cm
$x_0 = t_0 = 0$ $y_0 = 2.0$ m
$v_{0y} = 0$
$y_1 = 0$

Find
x_1

At top of circle

Solve: For the circular motion, Newton's second law along the r-direction is

$$\Sigma F_r = T + F_G = \frac{mv_t^2}{r}$$

Since the string goes slack as the particle makes it over the top, $T = 0$ N. That is,

$$F_G = mg = \frac{mv_t^2}{r} \Rightarrow v_t = \sqrt{gr} = \sqrt{(9.8 \text{ m/s}^2)(0.5 \text{ m})} = 2.21 \text{ m/s}$$

The ball begins projectile motion as the string is released. The time it takes for the ball to hit the floor can be found as follows:

$$y_1 = y_0 + v_{0y}(t_1 - t_0) + \tfrac{1}{2}a_y(t_1 - t_0)^2 \Rightarrow 0 \text{ m} = 2.0 \text{ m} + 0 \text{ m} + \tfrac{1}{2}(-9.8 \text{ m/s}^2)(t_1 - 0 \text{ s})^2 \Rightarrow t_1 = 0.639 \text{ s}$$

The place where the ball hits the ground is

$$x_1 = x_0 + v_{0x}(t_1 - t_0) = 0 \text{ m} + (+2.21 \text{ m/s})(0.639 \text{ s} - 0 \text{ s}) = +1.41 \text{ m}$$

The ball hits the ground 1.4 m to the right of the point beneath the center of the circle.

8.53. Model: Model the ball as a particle undergoing circular motion in a vertical circle.

Visualize:

Pictorial representation

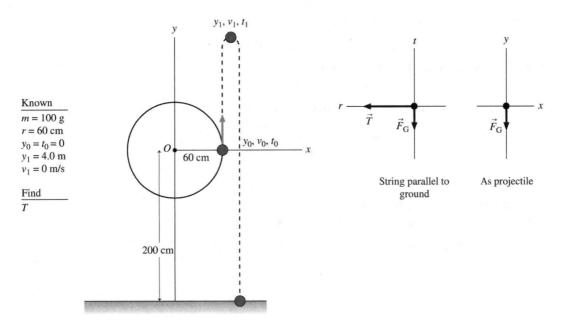

Known
$m = 100$ g
$r = 60$ cm
$y_0 = t_0 = 0$
$y_1 = 4.0$ m
$v_1 = 0$ m/s

Find
T

String parallel to ground   As projectile

Solve: Initially, the ball is moving in circular motion. Once the string breaks, it becomes a projectile. The final circular-motion velocity is the initial velocity for the projectile, which we can find by using the kinematic equation

$$v_1^2 = v_0^2 + 2a_y(y_1 - y_0) \Rightarrow 0 \text{ m}^2/\text{s}^2 = (v_0)^2 + 2(-9.8 \text{ m/s}^2)(4.0 \text{ m} - 0 \text{ m}) \Rightarrow v_0 = 8.85 \text{ m/s}$$

This is the speed of the ball as the string broke. The tension in the string at that instant can be found by using the r-component of the net force on the ball:

$$\sum F_r = T = m\left(\frac{v_{0y}^2}{r}\right) \Rightarrow T = (0.100 \text{ kg})\frac{(8.85 \text{ m/s})^2}{0.60 \text{ m}} = 13 \text{ N}$$

8.55. Model: Model the steel block as a particle and use the model of kinetic friction.
Visualize:

Pictorial representation

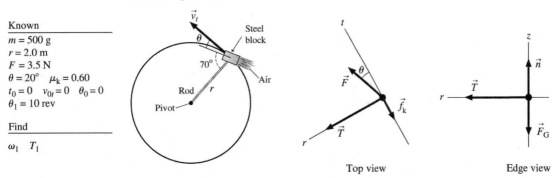

Known
$m = 500$ g
$r = 2.0$ m
$F = 3.5$ N
$\theta = 20°$ $\mu_k = 0.60$
$t_0 = 0$ $v_{0t} = 0$ $\theta_0 = 0$
$\theta_1 = 10$ rev

Find
ω_1 T_1

Top view   Edge view

Solve: **(a)** The components of thrust $(\vec{F})$ along the r-, t-, and z-directions are

$$F_r = F\sin 20° = (3.5 \text{ N})\sin 20° = 1.20 \text{ N} \qquad F_t = F\cos 20° = (3.5 \text{ N})\cos 20° = 3.29 \text{ N} \qquad F_z = 0 \text{ N}$$

Newton's second law is

$$(F_{net})_r = T + F_r = mr\omega^2 \quad (F_{net})_t = F_t - f_k = ma_t$$
$$(F_{net})_z = n - mg = 0 \text{ N}$$

The z-component equation means $n = mg$. The force of friction is

$$f_k = \mu_k n = \mu_k mg = (0.60)(0.500 \text{ kg})(9.8 \text{ m/s}^2) = 2.94 \text{ N}$$

Substituting into the t-component of Newton's second law

$$(3.29 \text{ N}) - (2.94 \text{ N}) = (0.500 \text{ kg})a_t \Rightarrow a_t = 0.70 \text{ m/s}^2$$

Having found a_t, we can now find the tangential velocity after 10 revolutions $= 20\pi$ rad as follows:

$$\theta_1 = \frac{1}{2}\left(\frac{a_t}{r}\right)t_1^2 \Rightarrow t_1 = \sqrt{\frac{2r\theta_1}{a_t}} = 18.95 \text{ s}$$

$$\omega_1 = \omega_0 + \left(\frac{a_t}{r}\right)t_1 = 6.63 \text{ rad/s}$$

The block's angular velocity after 10 rev is 6.6 rad/s.

(b) Substituting ω_1 into the r-component of Newton's second law yields:

$$T_1 + F_r = mr\omega_1^2 \Rightarrow T_1 + (1.20 \text{ N}) = (0.500 \text{ kg})(2.0 \text{ m})(6.63 \text{ rad/s})^2 \Rightarrow T_1 = 44 \text{ N}$$

IMPULSE AND MOMENTUM

Exercises and Problems

Section 9.1 Momentum and Impulse

9.1. Model: Model the car and the baseball as particles.
Solve: (a) The momentum $p = mv = (3000 \text{ kg})(15 \text{ m/s}) = 4.5 \times 10^4$ kg m/s.
(b) The momentum $p = mv = (0.20 \text{ kg})(40 \text{ m/s}) = 8.0$ kg m/s.

9.5. Visualize: Please refer to Figure EX9.5.
Solve: The impulse is defined in Equation 9.6 as $J_x = \int_{t_i}^{t_f} F_x(t)dt =$ area under the $F_x(t)$ curve between t_i and t_f.

For the force-time curve shown in Figure EX9.5, the impulse is $6.0 \text{ Ns} = \frac{1}{2}(F_{max})(8.0 \text{ ms}) \Rightarrow F_{max} = 1.5 \times 10^3$ N.

9.7. Model: Model the object as a particle and the interaction as a collision.
Visualize: Please refer to Figure EX9.7.
Solve: The object is initially moving to the right (positive momentum) and ends up moving to the left (negative momentum). Using the impulse-momentum theorem $p_{fx} = p_{ix} + J_x$,

$$-2 \text{ kg m/s} = +6 \text{ kg m/s} + J_x \quad \Rightarrow \quad J_x = -8 \text{ kg m/s} = -8 \text{ Ns}$$

Since $J_x = F_{avg}\Delta t$, we have

$$F_{avg}\Delta t = -8 \text{ Ns} \quad \Rightarrow \quad F_{avg} = \frac{-8 \text{ Ns}}{10 \text{ ms}} = -8 \times 10^2 \text{ N}$$

Thus, the force is 800 N to the left.

Section 9.2 Solving Impulse and Momentum Problems

9.9. Model: Model the object as a particle and the interaction with the force as a collision.
Visualize: Please refer to Figure EX9.9.
Solve: Using the equations

$$p_{fx} = p_{ix} + J_x \text{ and } J_x = \int_{t_i}^{t_f} F_x(t)dt = \text{area under force curve}$$

$$(2.0 \text{ kg})v_{fx} = (2.0 \text{ kg})(1.0 \text{ m/s}) + \text{(area under the force curve)}$$

$$v_{fx} = (1.0 \text{ m/s}) + \frac{1}{2.0 \text{ kg}}(-8.0 \text{ N})(0.50 \text{ s}) = -1.0 \text{ m/s}$$

Because v_{fx} is negative, the object is now moving to the left at 1.0 m/s.

Assess: The direction of the velocity has reversed.

9.13. Model: Model the glider cart as a particle, and its interaction with the spring as a collision. Assume a frictionless track.
Visualize:

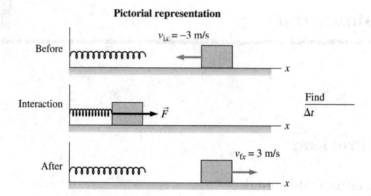

Pictorial representation

Solve: Using the impulse-momentum theorem $p_{fx} - p_{ix} = \int F(t)dt,$

$$(0.60 \text{ kg})(3 \text{ m/s}) - (0.60 \text{ kg})(-3 \text{ m/s}) = \text{area under force curve} = \tfrac{1}{2}(36 \text{ N})(\Delta t) \quad \Rightarrow \quad \Delta t = 0.2 \text{ s}$$

Section 9.4 Inelastic Collisions

9.17. Model: We will define our system to be bird + bug. This is the case of an inelastic collision because the bird and bug move together after the collision. Horizontal momentum is conserved because there are no external forces acting on the system during the collision in the impulse approximation.
Visualize:

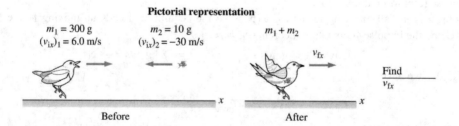

Pictorial representation

Solve: The conservation of momentum equation $p_{fx} = p_{ix}$ gives

$$(m_1 + m_2)v_{fx} = m_1(v_{ix})_1 + m_2(v_{ix})_2 \quad \Rightarrow \quad (300 \text{ g} + 10 \text{ g})v_{fx} = (300 \text{ g})(6.0 \text{ m/s}) + (10 \text{ g})(-30 \text{ m/s}) \quad \Rightarrow \quad v_{fx} = 4.8 \text{ m/s}$$

Assess: We left masses in grams, rather than convert to kilograms, because the mass units cancel out from both sides of the equation. Note that $(v_{ix})_2$ is negative because the bug is flying to the left.

9.19. Model: Because of external friction and drag forces, the car and the blob of sticky clay are not exactly an isolated system. But during the collision, friction and drag are not going to be significant. The momentum of the system will be conserved in the collision, within the impulse approximation.

Visualize:

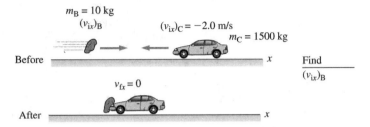

Pictorial representation

Solve: The conservation of momentum equation $p_{fx} = p_{ix}$ gives

$$(m_C + m_B)(v_f)_x = m_B(v_{ix})_B + m_C(v_{ix})_C$$

$$0 \text{ kg m/s} = (10 \text{ kg})(v_{ix})_B + (1500 \text{ kg})(-2.0 \text{ m/s}) \implies (v_{ix})_B = 3.0 \times 10^2 \text{ m/s}$$

Assess: This speed of the blob is around 600 mph, which is very large. However, a very large speed is *expected* in order to stop a car with only 10 kg of clay.

Section 9.5 Explosions

9.21. Model: We will define our system to be Dan + skateboard, and their interaction as an explosion. While friction is present between the skateboard and the ground, it is negligible in the impulse approximation.
Visualize:

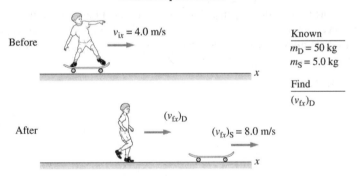

Pictorial representation

The system has nonzero initial momentum p_{ix}. As Dan (D) jumps backward off the gliding skateboard (S), the skateboard will move forward so that the final total momentum of the system p_{fx} is equal to p_{ix}.
Solve: We have $m_S(v_{fx})_S + m_D(v_{fx})_D = (m_S + m_D)v_{ix}$. Thus,

$$(5.0 \text{ kg})(8.0 \text{ m/s}) + (50 \text{ kg})(v_{fx})_D = (5.0 \text{ kg} + 50 \text{ kg})(4.0 \text{ m/s}) \implies (v_{fx})_D = 3.6 \text{ m/s}$$

Section 9.6 Momentum in Two Dimensions

9.25. Model: This problem deals with the conservation of momentum in two dimensions in an inelastic collision.

Visualize:

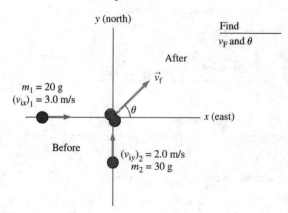

Pictorial representation

Solve: The conservation of momentum equation $\vec{p}_{\text{before}} = \vec{p}_{\text{after}}$ gives

$$m_1(v_{ix})_1 + m_2(v_{ix})_2 = (m_1 + m_2)v_{fx}, \quad m_1(v_{iy})_1 + m_2(v_{iy})_2 = (m_1 + m_2)v_{fy}$$

Substituting in the given values, we find

$$(0.020 \text{ kg})(3.0 \text{ m/s}) + 0.0 \text{ kg m/s} = (0.020 \text{ kg} + 0.030 \text{ kg})v_f \cos\theta$$

$$0.0 \text{ kg m/s} + (0.030 \text{ kg})(2.0 \text{ m/s}) = (0.020 \text{ kg} + 0.030 \text{ kg})v_f \sin\theta$$

$$v_f \cos\theta = 1.2 \text{ m/s}, \quad v_f \sin\theta = 1.2 \text{ m/s}$$

$$v_f = \sqrt{(1.2 \text{ m/s})^2 + (1.2 \text{ m/s})^2} = 1.7 \text{ m/s}, \quad \theta = \tan^{-1}\frac{v_y}{v_x} = \tan^{-1}(1) = 45°$$

The ball of clay moves 45° north of east at 1.7 m/s.

9.29. Model: Model the cart as a particle sliding down a frictionless ramp. The cart is subjected to an impulsive force when it comes in contact with a rubber block at the bottom of the ramp. We shall use the impulse-momentum theorem and the constant-acceleration kinematic equations.

Visualize:

Pictorial representation

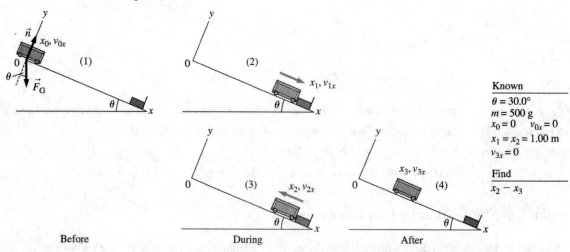

Known
$\theta = 30.0°$
$m = 500 \text{ g}$
$x_0 = 0 \quad v_{0x} = 0$
$x_1 = x_2 = 1.00 \text{ m}$
$v_{3x} = 0$

Find
$x_2 - x_3$

Solve: From the free-body diagram on the cart, Newton's second law applied to the system before the collision gives

$$\sum(F)_x = F_G \sin\theta = ma_x \quad \Rightarrow \quad a_x = \frac{mg \sin\theta}{m} = g \sin 30.0° = \frac{9.81 \text{ m/s}^2}{2} = 4.905 \text{ m/s}^2$$

Using this acceleration, we can find the cart's speed just before its contact with the rubber block:

$$v_{1x}^2 = v_{0x}^2 + 2a_x(x_1 - x_0) = 0 \text{ m}^2/\text{s}^2 + 2(4.905 \text{ m/s}^2)(1.00 \text{ m} - 0 \text{ m}) \implies v_{1x} = 3.132 \text{ m/s}$$

Now we can use the impulse-momentum theorem to obtain the velocity just after the collision:

$$mv_{2x} = mv_{1x} + \int F_x dt = mv_{1x} + \text{area under the force graph}$$

$$(0.500 \text{ kg})v_{2x} = (0.500 \text{ kg})(3.13 \text{ m/s}) - \tfrac{1}{2}(200 \text{ N})(26.7 \times 10^{-3} \text{ s}) \implies v_{2x} = -2.208 \text{ m/s}$$

Note that the given force graph is positive, but in this coordinate system the impulse of the force is to the left (i.e., up the slope). That is the reason to put a minus sign while evaluating the $\int F_x dt$ integral.

We can once again use a kinematic equation to find how far the cart will roll back up the ramp:

$$v_{3x}^2 = v_{2x}^2 + 2a_x(x_3 - x_2) \implies (0 \text{ m/s})^2 = (-2.208 \text{ m/s})^2 + 2(-4.905 \text{ m/s}^2)(x_3 - x_2) \implies (x_3 - x_2) = 0.497 \text{ m}$$

9.31. Solve: Apply Equation 9.7 and Newton's second law. The latter tells us that the average force used to expel the grains is

$$F_{avg} = ma = (1.0 \times 10^{-10} \text{ kg})(2.5 \times 10^4 \text{ m/s}^2) = 2.5 \times 10^{-6} \text{ N}$$

Inserting this into Equation 9.7 gives

$$J = F_{avg}\Delta t = (2.5 \times 10^{-6} \text{ N})(3.0 \times 10^4 \text{ s}) = 7.5 \times 10^{-10} \text{ kg m/s}$$

9.33. Visualize:

Pictorial representation

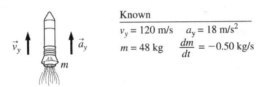

Solve: Using Newton's second law for the y-direction and the chain rule,

$$(F_{net})_y = \frac{dp_y}{dt} = \frac{d}{dt}(mv_y) = \frac{dm}{dt}(v_y) + m\left(\frac{dv_y}{dt}\right)$$

$$= (-0.50 \text{ kg/s})(120 \text{ m/s}) + (48 \text{ kg})(18 \text{ m/s}^2)$$

$$= 8.0 \times 10^2 \text{ N}$$

Assess: Since the rocket is losing mass, $dm/dt < 0$. The time derivative of the velocity is the acceleration.

9.39. Model: Model the squid and the water ejected as particles and ignore drag forces during the short time interval over which the water is expelled (the impulse approximation). Because the external forces are negligible, momentum will be conserved.
Visualize:

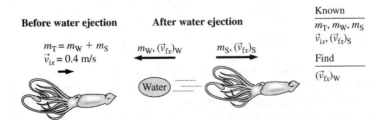

Solve: Applying conservation of momentum gives

$$m_T v_{1x} = m_W (v_{fx})_W + m_S (v_{fx})_S$$

$$(v_{fx})_W = \frac{1}{m_w}[m_T v_{1x} - m_S (v_{fx})_S] = \frac{1}{0.10 \text{ kg}}[(1.6 \text{ kg})(0.4 \text{ m/s}) - (1.5 \text{ kg})(2.5 \text{ m/s})] = -31.1 \text{ m/s}$$

This water is ejected in the direction opposite the squid's initial velocity, so the speed with which the water is ejected relative to the squid is

$$v_{rel} = (v_{fx})_W - (v_{fx})_S = -31.1 \text{ m/s} - 0.4 \text{ m/s} = -31.5 \text{ m/s}$$

or 32 m/s to two significant figures.

9.41. Model: This problem deals with a case that is the opposite of a collision. Our system is comprised of three coconut pieces that are modeled as particles. During the explosion, the total momentum of the system is conserved in the x- and y-directions.

Visualize:

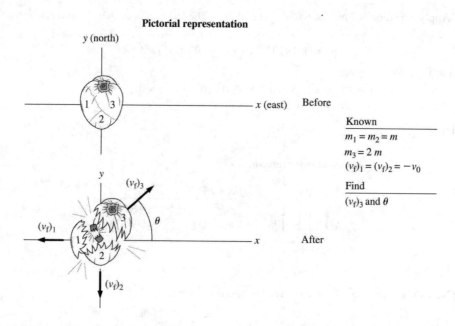

Solve: The initial momentum is zero. From $p_{fx} = p_{ix}$, we get

$$+m_1 (v_f)_1 + m_3 (v_f)_3 \cos\theta = 0 \text{ kg m/s} \quad \Rightarrow \quad (v_f)_3 \cos\theta = \frac{-m_1 (v_f)_1}{m_3} = \frac{-m(-v_0)}{2m} = \frac{v_0}{2}$$

From $p_{fx} = p_{ix}$, we get

$$+m_2 (v_f)_2 + m_3 (v_f)_3 \sin\theta = 0 \text{ kg m/s} \quad \Rightarrow \quad (v_f)_3 \sin\theta = \frac{-m_2 (v_f)_2}{m_3} = \frac{-m(-v_0)}{2m} = \frac{v_0}{2}$$

$$(v_f)_3 = \sqrt{\left(\frac{v_0}{2}\right)^2 + \left(\frac{v_0}{2}\right)^2} = \frac{v_0}{\sqrt{2}}, \ \theta = \tan^{-1}(1) = 45°$$

The speed to the third piece is 14 m/s at 45° east of north.

9.43. Model: Model the bullet and block as particles. This is an isolated system because any frictional force during the brief collision period is going to be insignificant. Within the impulse approximation, the momentum of our system will be conserved in the collision. After the collision, we will consider the frictional force and apply Newton's second law and kinematic equations to find the distance traveled by the block + bullet.

Visualize:

Pictorial representation

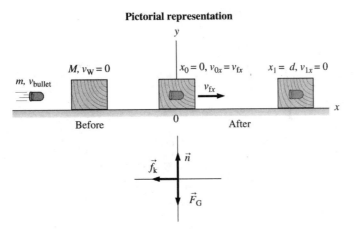

Solve: **(a)** Applying conservation of momentum to the collision gives

$$mv_{\text{bullet}} + Mv_W = (m + M)v_{\text{fx}} \quad \Rightarrow \quad v_{\text{bullet}} = \frac{m + M}{m}v_{\text{fx}}$$

The speed v_{ex} can be found from the kinematics equation

$$v_{1x}^2 = v_{0x}^2 + 2ad = v_{\text{fx}}^2 + 2ad \quad \Rightarrow \quad v_{\text{fx}} = \sqrt{-2ad}$$

The acceleration in the x-direction may be found using Newton's second law and the friction model. Because the block does not accelerate in the y-direction, the normal force must be the same magnitude as the force due to gravity (Newton's second law). Thus, the frictional force is $f_k = -\mu_k n = -\mu_k(m + M)g$, where the negative sign indicates that the force acts in the negative x-direction. Newton's second law then gives the acceleration of the block as

$$a = F_{\text{net}}/(m + M) = -\mu_k(m + M)g/(m + M) = -\mu_k g$$

Inserting this into the expression for v_{ex} gives

$$v_{\text{fx}} = \sqrt{-2ad} = \sqrt{2\mu_k gd}$$

Finally, we insert this expression for v_{ex} into the expression for the bullet's velocity to find

$$v_{\text{bullet}} = \frac{m + M}{m}\sqrt{2\mu_k gd}$$

(b) Inserting the given quantities gives

$$v_{\text{bullet}} = \frac{0.010\ \text{kg} + 10\ \text{kg}}{0.010\ \text{kg}}\sqrt{2(0.20)(9.8\ \text{m/s}^2)(0.050\ \text{m})} = 4.4 \times 10^2\ \text{m/s}$$

Visualize: If we let the bullet's mass go to zero, we see that the bullet's speed goes to infinity, which is reasonable because a zero-mass bullet would need an infinite speed to make the block move. If the bullet's mass goes to infinity, the bullet's speed would go to $\sqrt{2\mu_k gd}$, which is just the result for the initial speed of an object that decelerates to a stop at a constant rate $(\mu_k g)$ over a distance d. In other words, the block becomes insignificant compared to the infinite-mass bullet.

9.45. Model: Model the package and the rocket as particles. This is a two-part problem. First we have an inelastic collision between the rocket (R) and the package (P). During the collision, momentum is conserved since no significant external force acts on the rocket and the package. However, as soon as the package + rocket system leaves the cliff they become a projectile motion problem.

Visualize:

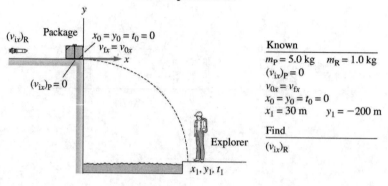

Pictorial representation

Solve: The minimum velocity after collision that the package + rocket must have to reach the explorer is v_{0x}, which can be found as follows:

$$y_1 = y_0 + v_{0y}(t_1 - t_0) + \tfrac{1}{2}a_y(t_1 - t_0)^2 \quad \Rightarrow \quad -200\,\text{m} = 0\,\text{m} + 0\,\text{m} + \tfrac{1}{2}(-9.8\,\text{m/s}^2)t_1^2 \quad \Rightarrow \quad t_1 = 6.389\,\text{s}$$

With this time, we can now find v_{0x} using $x_1 = x_0 + v_{0x}(t_1 - t_0) + \tfrac{1}{2}a_x(t_1 - t_0)^2$. We obtain

$$30\,\text{m} = 0\,\text{m} + v_{0x}(6.389\,\text{s}) + 0\,\text{m} \quad \Rightarrow \quad v_{0x} = 4.696\,\text{m/s} = v_{\text{fx}}$$

We now use the momentum conservation equation $p_{\text{fx}} = p_{\text{ix}}$ which can be written

$$(m_\text{R} + m_\text{P})v_{\text{fx}} = m_\text{R}(v_{\text{ix}})_\text{R} + m_\text{P}(v_{\text{ix}})_\text{P}$$
$$(1.0\,\text{kg} + 5.0\,\text{kg})(4.696\,\text{m/s}) = (1.0\,\text{kg})(v_{\text{ix}})_\text{R} + (5.0\,\text{kg})(0\,\text{m/s}) \quad \Rightarrow \quad (v_{\text{ix}})_\text{R} = 28\,\text{m/s}$$

9.51. Model: This is an inelastic collision. The total momentum of the Volkswagen + Cadillac system is conserved.
Visualize:

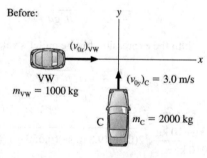

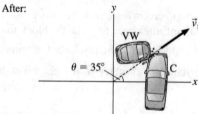

Find: $(v_{0x})_{\text{VW}}$

Solve: Apply conservation of momentum in the x- and y-directions.

$$(m_\text{C} + m_{\text{VW}})v_{1x} = (m_\text{C} + m_{\text{VW}})v_1\cos\theta = m_{\text{VW}}(v_{0x})_{\text{VW}}$$
$$(m_\text{C} + m_{\text{VW}})v_{1y} = (m_\text{C} + m_{\text{VW}})v_1\sin\theta = m_\text{C}(v_{0y})_\text{C}$$

From the y-equation, we find

$$v_1 = \frac{m_C(v_{0y})_C}{(m_C + m_{VW})\sin\theta} = \frac{(2000 \text{ kg})(3.0 \text{ m/s})}{(3000 \text{ kg})\sin 35°} = 3.49 \text{ m/s}$$

Inserting this value into the x-equation gives

$$(v_{0x})_{VW} = \frac{(m_C + m_{VW})v_1 \cos\theta}{m_{VW}} = \frac{(3000 \text{ kg})(3.49 \text{ m/s})\cos 35°}{1000 \text{ kg}} = 8.6 \text{ m/s}$$

9.53. Model: Assume that the tube is frictionless and ignore air resistance. Model the two balls as particles.
Visualize:

Solve: The initial momentum of the system is zero because both balls are stationary. Therefore, conservation of momentum tells us that the final momentum of the system must be zero:

$$mv_m + 3mv_0 = 0 \quad \Rightarrow \quad v_m = -3v_0$$

Thus, the speed of the lighter ball is $3v_0$.

Assess: The negative sign indicates that the lighter ball moves in the direction opposite the larger ball.

9.55. Solve: Apply the impulse-momentum theorem (Equation 9.8) to find the initial velocity:

$$\Delta p = \int F \, dt$$

$$v_{fx} = v_{ix} + \frac{1}{m}\int F \, dt$$

$$= -5.0 \text{ m/s} + \frac{1}{0.500 \text{ kg}} \int_{-2}^{2} (4 - t^2) \, dt$$

$$= -5.0 \text{ m/s} + \frac{1}{0.500 \text{ kg}} \left(4t - \frac{1}{3}t^3 \right)\Big|_{-2}^{2}$$

$$= -5.0 \text{ m/s} + \frac{1}{0.500 \text{ kg}} \left[8 - \frac{8}{3} - \left(-8 + \frac{8}{3} \right) \right]$$

$$= 16 \text{ m/s}$$

9.57. Model: This is a three-part problem. In the first part, the shell, treated as a particle, is launched as a projectile and reaches its highest point. We will use constant-acceleration kinematic equations for this part. The shell, which is our system, then explodes at the highest point. During this brief explosion time, momentum is conserved. In the third part, we will again use the kinematic equations to find the horizontal distance between the landing of the lighter fragment and the origin.

Visualize:

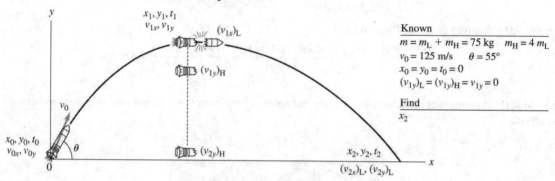

Pictorial representation

Solve: The initial velocity is

$$v_{0x} = v\cos\theta = (125 \text{ m/s})\cos 55° = 71.7 \text{ m/s}$$

$$v_{0y} = v\sin\theta = (125 \text{ m/s})\sin 55° = 102.4 \text{ m/s}$$

At the highest point, $v_{1y} = 0$ m/s and $v_{1x} = 71.7$ m/s. The conservation of momentum equation $p_{fx} = p_{ix}$ gives

$$m_L(v_{1x})_L + m_H(v_{1x})_H = (m_L + m_H)v_{1x}$$

The heavier particle falls straight down, so $(v_{1x})_H = 0$ m/s. Thus,

$$(15 \text{ kg})(v_{1x})_L + 0 \text{ kg m/s} = (15 \text{ kg} + 60 \text{ kg})(71.7 \text{ m/s}) \Rightarrow (v_{1x})_L = 358 \text{ m/s}$$

That is, the velocity of the smaller fragment immediately after the explosion is 358 m/s and this velocity is in the horizontal x-direction. Note that $(v_{1y})_L = 0$ m/s. To find x_2, we will first find the displacement $x_1 - x_0$ and then $x_2 - x_1$. For $x_1 - x_0$,

$$v_{1y} = v_{0y} + a_y(t_1 - t_0) \Rightarrow 0 \text{ m/s} = (102.4 \text{ m/s}) + (-9.8 \text{ m/s}^2)(t_1 - 0 \text{ s}) \Rightarrow t_1 = 10.45 \text{ s}$$

$$x_1 = x_0 + v_{0x}(t_1 - t_0) + \tfrac{1}{2}a_x(t_1 - t_0)^2 \Rightarrow x_1 - x_0 = (71.7 \text{ m/s})(10.45 \text{ s}) + 0 \text{ m} = 749 \text{ m}$$

For $x_2 - x_1$:

$$x_2 = x_1 + (v_{1x})_L(t_2 - t_1) + \tfrac{1}{2}a_x(t_2 - t_1)^2 \Rightarrow x_2 - x_1 = (358 \text{ m/s})(10.45 \text{ s}) + 0 \text{ m} = 3741 \text{ m}$$

$$x_2 = (x_2 - x_1) + (x_1 - x_0) = 3741 \text{ m} + 749 \text{ m} = 4490 \text{ m} = 4.5 \text{ km}$$

Assess: Note that the time of ascent to the highest point is equal to the time of descent to the ground, that is, $t_1 - t_0 = t_2 - t_1$.

9.63. Model: Model the three balls of clay as particle 1 (moving north), particle 2 (moving west), and particle 3 (moving southeast). The three stick together during their collision, which is perfectly inelastic. The momentum of the system is conserved.

Visualize:

Pictorial representation

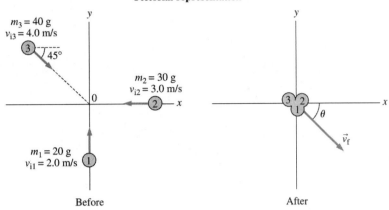

Before   After

Solve: The three initial momenta are

$$\vec{p}_{i1} = m_1\vec{v}_{i1} = (0.020 \text{ kg})(2.0 \text{ m/s})\hat{j} = 0.040\hat{j} \text{ kg m/s}$$

$$\vec{p}_{i2} = m_2\vec{v}_{i2} = (0.030 \text{ kg})(-3.0 \text{ m/s } \hat{i}) = -0.090\hat{i} \text{ kg m/s}$$

$$\vec{p}_{i3} = m_3\vec{v}_{i3} = (0.040 \text{ kg})[(4.0 \text{ m/s})\cos 45°\hat{i} - (4.0 \text{ m/s})\sin 45°\hat{j}] = (0.113\hat{i} - 0.113\hat{j}) \text{ kg m/s}$$

Since $\vec{p}_f = \vec{p}_i = \vec{p}_{i1} + \vec{p}_{i2} + \vec{p}_{i3}$, we have

$$(m_1 + m_2 + m_3)\vec{v}_f = (0.023\hat{i} - 0.073\hat{j}) \text{ kg m/s} \quad \Rightarrow \quad \vec{v}_f = (0.256\hat{i} - 0.811\hat{j}) \text{ m/s}$$

$$v_f = \sqrt{(0.256 \text{ m/s})^2 + (-0.811 \text{ m/s})^2} = 0.85 \text{ m/s}$$

$$\theta = \tan^{-1}\frac{|v_{fy}|}{v_{fx}} = \tan^{-1}\frac{0.811}{0.256} = 72° \text{ below the } x\text{-axis.}$$

9.65. Model: The ^{14}C atom undergoes an "explosion" and decays into a nucleus, an electron, and a neutrino. Momentum is conserved in the process of "explosion" or decay.
Visualize:

Pictorial representation

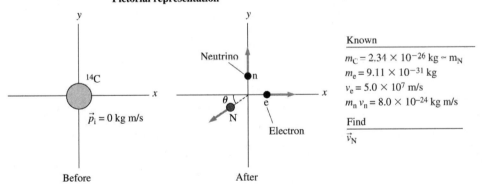

Before   After

Solve: The conservation of momentum equation $\vec{p}_f = \vec{p}_i = 0$ kg m/s gives

$$\vec{p}_e + \vec{p}_n + \vec{p}_N = 0 \text{ N} \quad \Rightarrow \quad \vec{p}_N = -(\vec{p}_e + \vec{p}_n) = -m_e\vec{v}_e - m_n\vec{v}_n$$

$$= -(9.11\times10^{-31} \text{ kg})(5.0\times10^7 \text{ m/s})\hat{i} - (8.0\times10^{-24} \text{ kg m/s})\hat{j} = -(45.55\times10^{-24}\hat{i} + 8.0\times10^{-24}\hat{j}) \text{ kg m/s}$$

$$p_N = m_N v_N = \sqrt{(45.55\times10^{-24})^2 + (8.0\times10^{-24})^2} \text{ kg m/s}$$

$$(2.34\times10^{-26} \text{ kg})v_N = 4.62\times10^{-23} \text{ kg m/s} \quad \Rightarrow \quad v_N = 2.0\times10^3 \text{ m/s}$$

ENERGY

Exercises and Problems

Section 10.2 Kinetic Energy and Gravitational Potential Energy

10.3. Model: Model the compact car (C) and the truck (T) as particles.
Visualize:

$m_C = 1000$ kg
v_C

$m_T = 20,000$ kg
$v_T = 25$ km/h

Solve: For the kinetic energy of the compact car and the kinetic energy of the truck to be equal,

$$K_C = K_T \Rightarrow \frac{1}{2}m_C v_C^2 = \frac{1}{2}m_T v_T^2 \Rightarrow v_C = \sqrt{\frac{m_T}{m_C}} v_T = \sqrt{\frac{20,000 \text{ kg}}{1000 \text{ kg}}} (25 \text{ km/h}) = 112 \text{ km/h}$$

Assess: A smaller mass needs a greater velocity for its kinetic energy to be the same as that of a larger mass.

10.5. Model: This is a case of free fall, so the sum of the kinetic and gravitational potential energy does not change as the ball rises and falls.
Visualize:

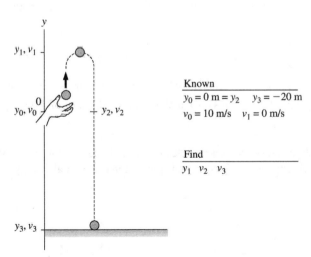

y

y_1, v_1

0

y_0, v_0

y_2, v_2

Known
$y_0 = 0$ m $= y_2$ $y_3 = -20$ m
$v_0 = 10$ m/s $v_1 = 0$ m/s

Find
y_1 v_2 v_3

y_3, v_3

The figure shows a ball's before-and-after pictorial representation for the three situations in parts (a), (b), and (c).

Solve: The quantity $K + U_g$ is the same during free fall: $K_f + U_{gf} = K_i + U_{gi}$. We have

(a) $\frac{1}{2}mv_1^2 + mgy_1 = \frac{1}{2}mv_0^2 + mgy_0$

$$\Rightarrow y_1 = (v_0^2 - v_1^2)/2g = [(10 \text{ m/s})^2 - (0 \text{ m/s})^2]/(2 \times 9.8 \text{ m/s}^2) = 5.10 \text{ m}$$

5.1 m is therefore the maximum height of the ball above the window. This is 25.1 m above the ground.

(b) $\frac{1}{2}mv_2^2 + mgy_2 = \frac{1}{2}mv_0^2 + mgy_0$

Since $y_2 = y_0 = 0$, we get for the magnitudes $v_2 = v_0 = 10 \text{ m/s}$.

(c) $\frac{1}{2}mv_3^2 + mgy_3 = \frac{1}{2}mv_0^2 + mgy_0 \Rightarrow v_3^2 + 2gy_3 = v_0^2 + 2gy_0 \Rightarrow v_3^2 = v_0^2 + 2g(y_0 - y_3)$

$$\Rightarrow v_3^2 = (10 \text{ m/s})^2 + 2(9.8 \text{ m/s}^2)[0 \text{ m} - (-20 \text{ m})] = 492 \text{ m}^2/\text{s}^2$$

This means the magnitude of v_3 is equal to 22 m/s.

Assess: Note that the ball's speed as it passes the window on its way down is the same as the speed with which it was tossed up, but in the opposite direction.

Section 10.3 A Closer Look at Gravitational Potential Energy

10.9. Model: Model the puck as a particle. Since the ramp is frictionless, the sum of the puck's kinetic and gravitational potential energy does not change during its sliding motion.

Visualize:

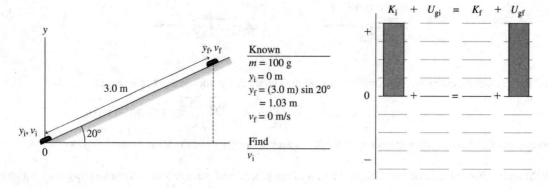

Solve: The quantity $K + U_g$ is the same at the top of the ramp as it was at the bottom. The energy conservation equation $K_f + U_{gf} = K_i + U_{gi}$ is

$$\frac{1}{2}mv_f^2 + mgy_f = \frac{1}{2}mv_i^2 + mgy_i \Rightarrow v_i^2 = v_f^2 + 2g(y_f - y_i)$$

$$\Rightarrow v_i^2 = (0 \text{ m/s})^2 + 2(9.8 \text{ m/s}^2)(1.03 \text{ m} - 0 \text{ m}) = 20.2 \text{ m}^2/\text{s}^2 \Rightarrow v_i = 4.5 \text{ m/s}$$

Assess: An initial push with a speed of 4.5 m/s ≈ 10 mph to cover a distance of 3.0 m up a 20° ramp seems reasonable.

10.11. Model: Model the child and swing as a particle, and assume the chain to be massless. In the absence of frictional and air-drag effects, the sum of the kinetic and gravitational potential energy does not change during the swing's motion.

Visualize:

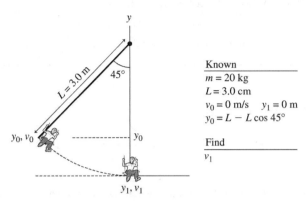

Known
$m = 20$ kg
$L = 3.0$ cm
$v_0 = 0$ m/s $y_1 = 0$ m
$y_0 = L - L \cos 45°$

Find

v_1

Solve: The quantity $K + U_g$ is the same at the highest point of the swing as it is at the lowest point. That is, $K_0 + U_{g0} = K_1 + U_{g1}$. It is clear from this equation that maximum kinetic energy occurs where the gravitational potential energy is the least. This is the case at the lowest position of the swing. At this position, the speed of the swing and child will also be maximum. The above equation is

$$\frac{1}{2}mv_0^2 + mgy_0 = \frac{1}{2}mv_1^2 + mgy_1 \Rightarrow v_1^2 = v_0^2 + 2g(y_0 - y_1)$$

$$\Rightarrow v_1^2 = (0 \text{ m/s})^2 + 2g(y_0 - 0 \text{ m}) \Rightarrow v_1 = \sqrt{2gy_0}$$

We see from the pictorial representation that

$$y_0 = L - L\cos 45° = (3.0 \text{ m}) - (3.0 \text{ m})\cos 45° = 0.879 \text{ m}$$

$$\Rightarrow v_1 = \sqrt{2gy_0} = \sqrt{2(9.8 \text{ m/s}^2)(0.879 \text{ m})} = 4.2 \text{ m/s}$$

Assess: We did not need to know the swing's or the child's mass. Also, a maximum speed of 4.2 m/s is reasonable.

Section 10.4 Restoring Forces and Hooke's Law

10.13. Model: Assume that the spring is ideal and obeys Hooke's law.
Visualize: According to Hooke's law, the spring force acting on a mass (m) attached to the end of a spring is given as $F_{sp} = k\Delta x$, where Δx is the change in length of the spring. If the mass m is at rest, then F_{sp} is also equal to the gravitational force $F_G = mg$.
Solve: We have $F_{sp} = k\Delta x = mg$. We want a 0.100 kg mass to give $\Delta x = 0.010$ m. This means

$$k = mg/\Delta x = (0.100 \text{ kg})(9.8 \text{ N/m})/(0.010 \text{ m}) = 98 \text{ N/m}$$

10.15. Model: Model the student (S) as a particle and the spring as obeying Hooke's law.
Visualize:

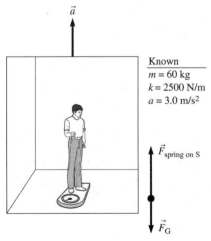

Known
$m = 60$ kg
$k = 2500$ N/m
$a = 3.0$ m/s^2

Solve: According to Newton's second law the force on the student is

$$\Sigma(F_{\text{on S}})_y = F_{\text{spring on S}} - F_G = ma_y$$

$$\Rightarrow F_{\text{spring on S}} = F_G + ma_y = mg + ma_y = (60 \text{ kg})(9.8 \text{ m/s}^2 + 3.0 \text{ m/s}^2) = 768 \text{ N}$$

Since $F_{\text{spring on S}} = F_{\text{S on spring}} = k\Delta y$, $k\Delta y = 768$ N. This means $\Delta y = (768 \text{ N})/(2500 \text{ N/m}) = 0.31$ m.

Section 10.5 Elastic Potential Energy

10.19. Model: Assume the spring is ideal and obeys Hooke's law. Then the potential energy of a stretched spring is $U_{\text{sp}} = \frac{1}{2}k(\Delta s)^2$.

Visualize: Use ratios to solve this problem. Use primed variables for the new situation with the spring stretched three times as far.

Solve:

$$\frac{U'_{\text{sp}}}{U_{\text{sp}}} = \frac{\frac{1}{2}k(\Delta s')^2}{\frac{1}{2}k(\Delta s)^2} = \frac{\frac{1}{2}k(3\Delta s)^2}{\frac{1}{2}k(\Delta s)^2} = 3^2 = 9$$

$$U'_{\text{sp}} = 9U_{\text{sp}} = 9(2.0 \text{ J}) = 18 \text{ J}$$

Assess: The stored energy scales with the square of the spring stretch.

10.21. Model: Assume an ideal spring that obeys Hooke's law. Since there is no friction, the mechanical energy $K + U_s$ is conserved. Also, model the block as a particle.

Visualize:

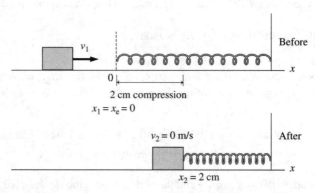

The figure shows a before-and-after pictorial representation. We have put the origin of our coordinate system at the equilibrium position of the free end of the spring. This gives us $x_1 = x_e = 0$ cm and $x_2 = 2.0$ cm.

Solve: The conservation of energy equation $K_2 + U_{s2} = K_1 + U_{s1}$ is

$$\frac{1}{2}mv_2^2 + \frac{1}{2}k(x_2 - x_e)^2 = \frac{1}{2}mv_1^2 + \frac{1}{2}k(x_1 - x_e)^2$$

Using $v_2 = 0$ m/s, $x_1 = x_e = 0$ m, and $x_2 - x_e = 0.020$ m, we get

$$\frac{1}{2}k(x_2 - x_e)^2 = \frac{1}{2}mv_1^2 \Rightarrow \Delta x = (x_2 - x_e) = \sqrt{\frac{m}{k}}v_1$$

That is, the compression is directly proportional to the velocity v_1. When the block collides with the spring with twice the earlier velocity $(2v_1)$, the compression will also be doubled to $2(x_2 - x_e) = 2(2.0 \text{ cm}) = 4.0$ cm.

Assess: This problem shows the power of using energy conservation over using Newton's laws in solving problems involving nonconstant acceleration.

Section 10.6 Energy Diagrams

10.25. Model: For an energy diagram, the sum of the kinetic and potential energy is a constant.
Visualize:

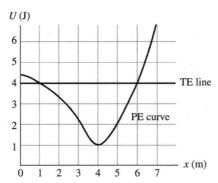

The particle is released from rest at $x = 1.0$ m. That is, $K = 0$ at $x = 1.0$ m. Since the total energy is given by $E = K + U$, we can draw a horizontal total energy (TE) line through the point of intersection of the potential energy curve (PE) and the $x = 1.0$ m line. The distance from the PE curve to the TE line is the particle's kinetic energy. These values are transformed as the position changes, causing the particle to speed up or slow down, but the sum $K + U$ does not change.

Solve: (a) We have $E = 4.0$ J and this energy is a constant. For $x < 1.0, U > 4.0$ J and, therefore, K must be negative to keep E the same (note that $K = E - U$ or $K = 4.0$ J $- U$). Since negative kinetic energy is unphysical, the particle cannot move to the left. That is, the particle will move to the right of $x = 1.0$ m.

(b) The expression for the kinetic energy is $E - U$. This means the particle has maximum speed or maximum kinetic energy when U is minimum. This happens at $x = 4.0$ m. Thus,

$$K_{max} = E - U_{min} = (4.0 \text{ J}) - (1.0 \text{ J}) = 3.0 \text{ J} \quad \frac{1}{2}mv_{max}^2 = 3.0 \text{ J} \Rightarrow v_{max} = \sqrt{\frac{2(3.0 \text{ J})}{m}} = \sqrt{\frac{8.0 \text{ J}}{0.020 \text{ kg}}} = 17.3 \text{ m/s}$$

The particle possesses this speed at $x = 4.0$ m.

(c) The total energy (TE) line intersects the potential energy (PE) curve at $x = 1.0$ m and $x = 6.0$ m. These are the turning points of the motion.

10.27. Model: For an energy diagram, the sum of the kinetic and potential energy is a constant.
Visualize:

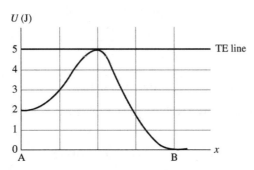

For the speed of the particle at A that is needed to reach B to be a minimum, the particle's kinetic energy as it reaches the top must be zero. Similarly, the minimum speed at B for the particle to reach A obtains when the particle just makes it to the top with zero kinetic energy.

Solve: (a) The energy equation $K_A + U_A = K_{top} + U_{top}$ is

$$\frac{1}{2}mv_A^2 + U_A = 0 \text{ J} + U_{top}$$

$$\Rightarrow v_A = \sqrt{2(U_{top} - U_A)/m} = \sqrt{2(5.0 \text{ J} - 2.0 \text{ J})/0.100 \text{ kg}} = 7.7 \text{ m/s}$$

(b) To go from point B to point A, $K_B + U_B = K_{top} + U_{top}$ is

$$\frac{1}{2}mv_B^2 + U_B = 0\ \text{J} + U_{top}$$

$$\Rightarrow v_B = \sqrt{2(U_{top} - U_B)/m} = \sqrt{2(5.0\ \text{J} - 0\ \text{J})/0.100\ \text{kg}} = 10.0\ \text{m/s}$$

Assess: The particle requires a higher kinetic energy to reach A from B than to reach B from A.

Section 10.7 Elastic Collisions

10.31. Model: In this case of a one-dimensional collision, the momentum conservation law is obeyed whether the collision is perfectly elastic or perfectly inelastic. Assume ball 1 is initially moving right, in the positive direction.
Visualize:

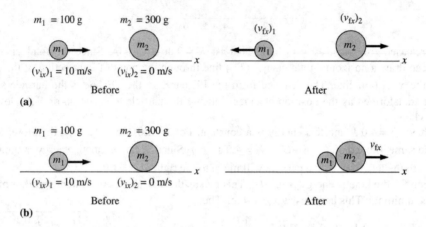

(a)

(b)

Solve: In the case of a perfectly elastic collision, the two velocities $(v_{fx})_1$ and $(v_{fx})_2$ can be determined by combining the conservation equations of momentum and mechanical energy. By contrast, a perfectly inelastic collision involves only one final velocity v_{fx} and can be determined from just the momentum conservation equation.

(a) Momentum conservation: $m_1(v_{ix})_1 + m_2(v_{ix})_2 = m_1(v_{fx})_1 + m_2(v_{fx})_2$

Energy conservation: $\frac{1}{2}m_1(v_{ix})_1^2 + \frac{1}{2}m_2(v_{ix})_2^2 = \frac{1}{2}m_1(v_{fx})_1^2 + \frac{1}{2}m_2(v_{fx})_2^2$

These two equations can be solved as shown in Equations 10.38 through 10.42:

$$(v_{fx})_1 = \frac{m_1 - m_2}{m_1 + m_2}(v_{ix})_1 = \frac{(100\ \text{g}) - (300\ \text{g})}{(100\ \text{g}) + (300\ \text{g})}(10\ \text{m/s}) = -5.0\ \text{m/s}$$

$$(v_{fx})_2 = \frac{2m_1}{m_1 + m_2}(v_{ix})_1 = \frac{2(100\ \text{g})}{(100\ \text{g}) + (300\ \text{g})}(10\ \text{m/s}) = +5.0\ \text{m/s}$$

(b) For the inelastic collision, both balls travel with the same final speed v_{fx}. The momentum conservation equation $p_{fx} = p_{ix}$ is

$$(m_1 + m_2)v_{fx} = m_1(v_{ix})_1 + m_2(v_{ix})_2$$

$$\Rightarrow v_{fx} = \left(\frac{100\ \text{g}}{100\ \text{g} + 300\ \text{g}}\right)(10\ \text{m/s}) + 0\ \text{m/s} = 2.5\ \text{m/s}$$

Assess: In the case of the perfectly elastic collision, the two balls bounce off each other with a speed of 5.0 m/s. In the case of the perfectly inelastic collision, the balls stick together and move together at 2.5 m/s.

10.33. Model: We will take the system to be the person plus the earth.

Visualize: When a person drops from a certain height, the initial potential energy is transformed to kinetic energy. When the person hits the ground, if they land rigidly upright, we assume that all of this energy is transformed into elastic potential energy of the compressed leg bones. The maximum energy that can be absorbed by the leg bones is 200 J; this limits the maximum height.

Solve: (a) The initial potential energy can be at most 200 J, so the height h of the jump is limited by $mgh = 200$ J

For $m = 60$ kg, this limits the height to

$$h = 200 \text{ J/mg} = 200 \text{ J/(60 kg)}(9.8 \text{ m/s}^2) = 0.34 \text{ m}$$

(b) If some of the energy is transformed to other forms than elastic energy in the bones, the initial height can be greater. If a person flexes her legs on landing, some energy is transformed to thermal energy. This allows for a greater initial height.

Assess: There are other tissues in the body with elastic properties that will absorb energy as well, so this limit is quite conservative.

10.37. Model: For the ice cube sliding around the inside of a smooth pipe, the sum of the kinetic and gravitational potential energy does not change.

Visualize:

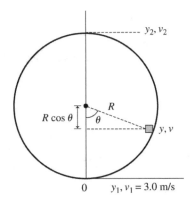

We use a coordinate system with the origin at the bottom of the pipe, that is, $y_1 = 0$. The radius R of the pipe is 10 cm, and therefore $y_{\text{top}} = y_2 = 2R = 0.20$ m. At an arbitrary angle θ, measured counterclockwise from the bottom of the circle, $y = R - R\cos\theta$.

Solve: (a) The energy conservation equation $K_2 + U_{g2} = K_1 + U_{g1}$ is

$$\Rightarrow \frac{1}{2}mv_2^2 + mgy_2 = \frac{1}{2}mv_1^2 + mgy_1$$

$$\Rightarrow v_2 = \sqrt{v_1^2 + 2g(y_1 - y_2)} = \sqrt{(3.0 \text{ m/s})^2 + 2(9.8 \text{ m/s}^2)(0 \text{ m} - 0.20 \text{ m})} = 2.25 \text{ m/s} \approx 2.3 \text{ m/s}$$

(b) Expressing the energy conservation equation as a function of θ:

$$K(\theta) + U_g(\theta) = K_1 + U_{g1} \Rightarrow \frac{1}{2}mv^2(\theta) + mgy(\theta) = \frac{1}{2}mv_1^2 + 0 \text{ J}$$

$$\Rightarrow v(\theta) = \sqrt{v_1^2 - 2gy(\theta)} = \sqrt{v_1^2 - 2gR(1 - \cos\theta)}$$

Using $v_1 = 3.0$ m/s, $g = 9.8$ m/s^2, and $R = 0.10$ m, we get $v(\theta) = \sqrt{9 - 1.96(1 - \cos\theta)}$ (m/s)

Assess: Beginning with a speed of 3.0 m/s at the bottom, the marble's potential energy increases and kinetic energy decreases as it gets toward the top of the circle. At the top, its speed is 2.25 m/s. This is reasonable since some of the kinetic energy has been transformed into the marble's potential energy.

10.39. Model: We will assume the knee extensor tendon behaves according to Hooke's Law and stretches in a straight line.

Visiualize: The elastic energy stored in a spring is given by $U_s = \frac{1}{2}k(\Delta s)^2$.

Solve: For athletes,

$$U_{s,\text{athlete}} = \frac{1}{2}k(\Delta s)^2 = \frac{1}{2}(33{,}000 \text{ N/m})(0.041 \text{ m})^2 = 27.7 \text{ J}$$

For non-athletes,

$$U_{s,\text{non-athlete}} = \frac{1}{2}k(\Delta s)^2 = \frac{1}{2}(33{,}000 \text{ N/m})(0.033 \text{ m})^2 = 18.0 \text{ J}$$

The difference in energy stored between athletes and non-athletes is therefore 9.7 J.

Assess: Notice the energy stored by athletes is over 1.5 times the energy stored by non-athletes.

10.43. Model: Model the two packages as particles. Momentum is conserved in both inelastic and elastic collisions. Kinetic energy is conserved only in a perfectly elastic collision.

Visualize:

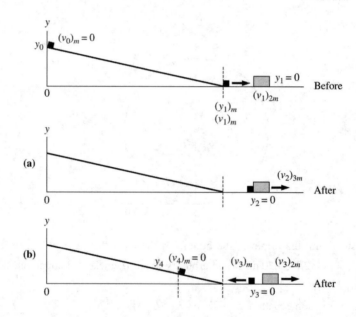

Solve: For a package with mass m the conservation of energy equation is

$$K_1 + U_{g1} = K_0 + U_{g0} \Rightarrow \frac{1}{2}m(v_1)_m^2 + mgy_1 = \frac{1}{2}m(v_0)_m^2 + mgy_0$$

Using $(v_0)_m = 0$ m/s and $y_1 = 0$ m,

$$\frac{1}{2}m(v_1)_m^2 = mgy_0 \Rightarrow (v_1)_m = \sqrt{2gy_0} = \sqrt{2(9.8 \text{ m/s}^2)(3.0 \text{ m})} = 7.668 \text{ m/s}$$

(a) For the perfectly inelastic collision the conservation of momentum equation is

$$p_{\text{fx}} = p_{\text{ix}} \Rightarrow (m + 2m)(v_2)_{3m} = m(v_1)_m + (2m)(v_1)_{2m}$$

Using $(v_1)_{2m} = 0$ m/s, we get

$$(v_2)_{3m} = (v_1)_m/3 = 2.56 \text{ m/s}$$

The packages move off together at a speed of 2.6 m/s.

(b) For the elastic collision, the mass m package rebounds with velocity

$$(v_3)_m = \frac{m-2m}{m+2m}(v_1)_m = -\frac{1}{3}(7.668 \text{ m/s}) = -2.56 \text{ m/s}$$

The negative sign with $(v_3)_m$ shows that the package with mass m rebounds and goes to the position y_4. We can determine y_4 by applying the conservation of energy equation as follows. For a package of mass m:

$$K_f + U_{gf} = K_i + U_{gi} \Rightarrow \frac{1}{2}m(v_4)_m^2 + mgy_4 = \frac{1}{2}m(v_3)_m^2 + mgy_3$$

Using $(v_3)_m = -2.55$ m/s, $y_3 = 0$ m, and $(v_4)_m = 0$ m/s, we get

$$mgy_4 = \frac{1}{2}m(-2.56 \text{ m/s})^2 \Rightarrow y_4 = 33 \text{ cm}$$

10.45. Model: Assume an ideal spring that obeys Hooke's law. Since this is a free-fall problem, the mechanical energy $K + U_g + U_s$ is conserved. Also, model the safe as a particle.

Visualize:

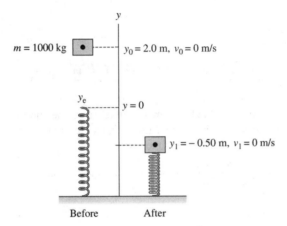

We have chosen to place the origin of our coordinate system at the free end of the spring, which is neither stretched nor compressed. The safe gains kinetic energy as it falls. The energy is then converted into elastic potential energy as the safe compresses the spring. The only two forces are gravity and the spring force, which are both conservative, so energy is conserved throughout the process. This means that the initial energy—as the safe is released—equals the final energy—when the safe is at rest and the spring is fully compressed.

Solve: The conservation of energy equation $K_1 + U_{g1} + U_{s1} = K_0 + U_{g0} + U_{s0}$ is

$$\frac{1}{2}mv_1^2 + mg(y_1 - y_e) + \frac{1}{2}k(y_1 - y_e)^2 = \frac{1}{2}mv_0^2 + mg(y_0 - y_e) + \frac{1}{2}k(y_e - y_e)^2$$

Using $v_1 = v_0 = 0$ m/s and $y_e = 0$ m, the above equation simplifies to

$$mgy_1 + \frac{1}{2}ky_1^2 = mgy_0$$

$$\Rightarrow k = \frac{2mg(y_0 - y_1)}{y_1^2} = \frac{2(1000 \text{ kg})(9.8 \text{ m/s}^2)(2.0 \text{ m} - (-0.50 \text{ m}))}{(-0.50 \text{ m})^2} = 1.96 \times 10^5 \text{ N/m} \approx 2.0 \times 10^5 \text{ N/m}$$

Assess: By equating energy at these two points, we do not need to find how fast the safe was moving when it hit the spring.

10.49. Model: We assume the spring to be ideal and to obey Hooke's law. We also treat the block (B) and the ball (b) as particles. In the case of an elastic collision, both the momentum and kinetic energy equations apply. On the other hand, for a perfectly inelastic collision only the equation of momentum conservation is valid.

Visualize:

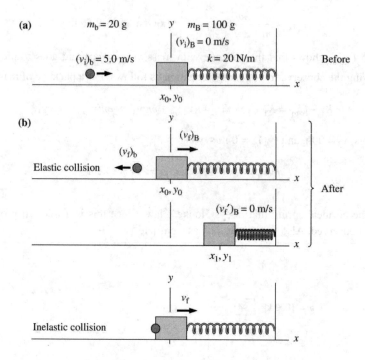

(a)

(b)

Place the origin of the coordinate system on the block that is attached to one end of the spring. The before-and-after pictorial representations of the elastic and perfectly inelastic collision are shown in figures (a) and (b), respectively.
Solve: **(a)** For an elastic collision, the ball's rebound velocity is

$$(v_f)_b = \frac{m_b - m_B}{m_b + m_B}(v_i)_b = \frac{-80 \text{ g}}{120 \text{ g}}(5.0 \text{ m/s}) = -3.33 \text{ m/s}$$

The ball's speed is 3.3 m/s.
(b) An elastic collision gives the block speed

$$(v_f)_B = \frac{2m_B}{m_b + m_B}(v_i)_b = \frac{40 \text{ g}}{120 \text{ g}}(5.0 \text{ m/s}) = 1.667 \text{ m/s}$$

To find the maximum compression of the spring, we use the conservation equation of mechanical energy for the block + spring system. That is $K_1 + U_{s1} = K_0 + U_{s0}$:

$$\frac{1}{2}m_B(v_f')_B^2 + \frac{1}{2}k(x_1 - x_0)^2 = \frac{1}{2}m_B(v_f)_B^2 + \frac{1}{2}k(x_0 - x_0)^2 \quad 0 + k(x_1 - x_0)^2 = m_B(v_f)_B^2 + 0$$

$$(x_1 - x_0) = \sqrt{(0.100 \text{ kg})(1.667 \text{ m/s})^2/(20 \text{ N/m})} = 11.8 \text{ cm}$$

(c) Momentum conservation $p_f = p_i$ for the perfectly inelastic collision means

$$(m_B + m_B)v_f = m_b(v_i)_b + m_B(v_i)_B$$

$$(0.100 \text{ kg} + 0.020 \text{ kg})v_f = (0.020 \text{ kg})(5.0 \text{ m/s}) + 0 \text{ m/v} \Rightarrow v_f = 0.833 \text{ m/s}$$

The maximum compression in this case can now be obtained using the conservation of energy equation $K_1 + U_{s1} = K_0 + U_{s0}$:

$$0 \text{ J} + (1/2)k(\Delta x)^2 = (1/2)(m_B + m_b)v_f^2 + 0 \text{ J}$$

$$\Rightarrow \Delta x = \sqrt{\frac{m_B + m_b}{k}}v_f = \sqrt{\frac{0.120 \text{ kg}}{20 \text{ N/m}}}(0.833 \text{ m/s}) = 0.0645 \text{ m} = 6.5 \text{ cm}$$

10.51. Model: Assume an ideal spring that obeys Hooke's law. There is no friction, hence the mechanical energy $K + U_g + U_s$ is conserved.

Visualize:

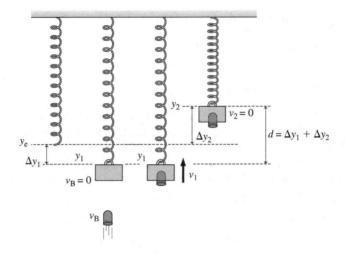

We have chosen to place the origin of the coordinate system on the free end of the spring that is neither stretched nor compressed, that is, at the equilibrium position of the end of the unstretched spring. The bullet's mass is m and the block's mass is M.

Solve: (a) The energy conservation equation $K_2 + U_{s2} + U_{g2} = K_1 + U_{s1} + U_{g1}$ becomes

$$\frac{1}{2}(m+M)v_2^2 + \frac{1}{2}k(y_2 - y_e)^2 + (m+M)g(y_2 - y_e) = \frac{1}{2}(m+M)v_1^2 + \frac{1}{2}k(y_1 - y_e)^2 - (m+M)g(y_1 - y_e)$$

Noting $v_2 = 0$ m/s, we can rewrite the above equation as

$$k(\Delta y_2)^2 + 2(m+M)g(\Delta y_2 + \Delta y_1) = (m+M)v_1^2 + k(\Delta y_1)^2$$

Let us express v_1 in terms of the bullet's initial speed v_B by using the momentum conservation equation $p_f = p_i$ which is $(m+M)v_1 = mv_B + Mv_{block}$. Since $v_{block} = 0$ m/s, we have

$$v_1 = \left(\frac{m}{m+M}\right)v_B$$

We can also find the magnitude of y_1 from the equilibrium condition $k(y_1 - y_e) = Mg$.

$$\Delta y_1 = \frac{Mg}{k}$$

With these substitutions for v_1 and Δy_1, the energy conservation equation simplifies to

$$k(\Delta y_2)^2 + 2(m+M)g(\Delta y_1 + \Delta y_2) = \frac{m^2 v_B^2}{(m+M)} + k\left(\frac{Mg}{k}\right)^2$$

$$\Rightarrow v_B^2 = 2\left(\frac{m+M}{m}\right)^2 g(\Delta y_1 + \Delta y_2) - \frac{(m+M)}{m^2}\frac{M^2 g^2}{k} + k\left(\frac{m+M}{m^2}\right)(\Delta y_2)^2$$

We still need to include the spring's maximum compression (d) into this equation. We assume that $d = \Delta y_1 + \Delta y_2$, that is, maximum compression is measured from the initial position (y_1) of the block. Thus, using $\Delta y_2 = d - \Delta y_1 = (d - Mg/k)$, we have

$$v_B = \left[2\left(\frac{m+M}{m}\right)^2 gd - \left(\frac{m+M}{m^2}\right)\frac{M^2 g^2}{k} + k\left(\frac{m+M}{m^2}\right)(d - Mg/k)^2\right]^{1/2}$$

(b) Using $m = 0.010$ kg, $M = 2.0$ kg, $k = 50$ N/m, and $d = 0.45$ m,

$$v_B^2 = 2\left(\frac{2.010 \text{ kg}}{0.010 \text{ kg}}\right)^2 (9.8 \text{ m/s}^2)(0.45 \text{ m}) - (2.010 \text{ kg})\left(\frac{2.0 \text{ kg}}{0.010 \text{ kg}}\right)^2 (9.8 \text{ m/s}^2)^2 / (50 \text{ N/m})$$

$$+ (50 \text{ N/m})(2.010 \text{ kg})\frac{1}{(0.010 \text{ kg})^2}[0.45 \text{ m} - (2.0 \text{ kg}) \times (9.8 \text{ m/s}^2)/50 \text{ N/m}]^2$$

$$\Rightarrow v_B = 453 \text{ m/s}$$

The bullet has a speed of 4.5×10^2 m/s.

Assess: This is a reasonable speed for the bullet.

10.57. Model: Model the balls as particles. We will use the Galilean transformation of velocities (Equation 10.43) to analyze the problem of elastic collisions. We will transform velocities from the lab frame L to a frame M in which one ball is at rest. This allows us to apply Equations 10.43 to a perfectly elastic collision in M. After finding the final velocities of the balls in M, we can then transform these velocities back to the lab frame L.

Visualize: Let M be the frame of the 400 g ball. Denoting masses as $m_1 = 100$ g and $m_2 = 400$ g, the initial velocities in the S frame are $(v_{ix})_{1L} = +4.0$ m/s and $(v_{ix})_{2L} = +1.0$ m/s.

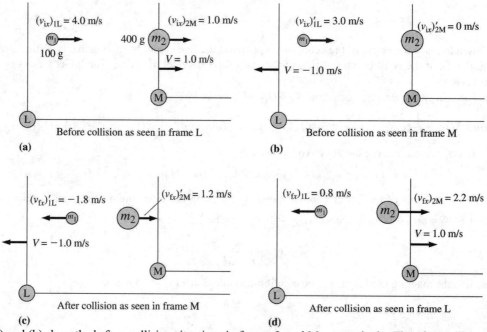

Figures (a) and (b) show the before-collision situations in frames L and M, respectively. The after-collision velocities in M are shown in figure (c). Figure (d) indicates velocities in L after they have been transformed to L from M.

Solve: In frame L, $(v_{ix})_{1L} = 4.0$ m/s and $(v_{ix})_{2L} = 1.0$ m/s. Because M is the reference frame of the 400 g ball, $(v_x)_{ML} = 1.0$ m/s. The velocities of the two balls in this frame can be obtained using the Galilean transformation of velocities $(v_x)_{OM} = (v_x)_{OL} - (v_x)_{ML}$. So,

$$(v_{ix})_{1M} = (v_{ix})_{1L} - (v_x)_{ML} = 4.0 \text{ m/s} - 1.0 \text{ m/s} = 3.0 \text{ m/s} \quad (v_{ix})_{2M} = (v_{ix})_{2L} - (v_x)_{ML} = 1.0 \text{ m/s} - 1.0 \text{ m/s} = 0 \text{ m/s}$$

Figure (b) shows the "before" situation in frame M where the ball 2 is at rest.

Now we can use Equations 10.43 to find the after-collision velocities in frame M.

$$(v_{fx})_{1M} = \frac{m_1 - m_2}{m_1 + m_2}(v_{ix})_{1M} = \frac{100 \text{ g} - 400 \text{ g}}{100 \text{ g} + 400 \text{ g}}(3.0 \text{ m/s}) = -1.80 \text{ m/s}$$

$$(v_{fx})_{2M} = \frac{2m_1}{m_1 + m_2}(v_{ix})_{1M} = \frac{2(100 \text{ g})}{100 \text{ g} + 400 \text{ g}}(3.0 \text{ m/s}) = 1.20 \text{ m/s}$$

Finally, we need to apply the reverse Galilean transformation $(v_x)_{OM} = (v_x)_{OL} + (v_x)_{LM}$ with the same $(v_x)_{LM}$, to transform the after-collision velocities back to the lab frame L.

$$(v_{fx})_{1L} = (v_{fx})_{1M} + (v_x)_{ML} = -1.80 \text{ m/s} + 1.0 \text{ m/s} = -0.80 \text{ m/s}$$
$$(v_{fx})_{2L} = (v_{fx})_{2M} + (v_x)_{ML} = 1.20 \text{ m/s} + 1.0 \text{ m/s} = 2.20 \text{ m/s}$$

Figure (d) shows the "after" situation in frame L. The 100 g ball moves left at 0.80 m/s, the 400 g ball right at 2.2 m/s.
Assess: The magnitudes of the after-collision velocities are similar to the magnitudes of the before-collision velocities.

10.59. Solve: (a) The equilibrium positions are located at points where $\dfrac{dU}{dx} = 0$.

$$\frac{dU}{dx} = 0 = 1 + 2\cos(2x) \Rightarrow \cos(2x) = -\frac{1}{2}$$
$$\Rightarrow x = \frac{1}{2}\cos^{-1}\left(-\frac{1}{2}\right)$$

Note that $-\dfrac{1}{2}$ is in radians and x is in meters. The function $\cos^{-1}\left(-\dfrac{1}{2}\right)$ may have values $\dfrac{2\pi}{3}$ and $\dfrac{4\pi}{3}$. Thus there are two values of x,

$$x_1 = \frac{\pi}{3} \text{ and } x_2 = \frac{2\pi}{3}$$

within the interval $0 \text{ m} \le x \le \pi \text{ m}$.
(b) A point of stable equilibrium corresponds to a local minimum, while a point of unstable equilibrium corresponds to a local maximum. Compute the concavity of $U(x)$ at the equilibrium positions to determine their stability.

$$\frac{d^2U}{dx^2} = -4\sin(2x)$$

At $x_1 = \dfrac{\pi}{3}$, $\dfrac{d^2U}{dx^2}(x_1) = -4\left(\dfrac{\sqrt{3}}{2}\right) = -2\sqrt{3}$. Since $\dfrac{d^2U}{dx^2}(x_1) < 0$, $x_1 = \dfrac{\pi}{3}$ is a local maximum, so $x_1 = \dfrac{\pi}{3}$ is a point of unstable equilibrium.

At $x_2 = \dfrac{2\pi}{3}$, $\dfrac{d^2U}{dx^2}(-x_2) = -4\left(-\dfrac{\sqrt{3}}{2}\right) = +2\sqrt{3}$. Since $\dfrac{d^2U}{dx^2} > 0$, $x_2 = \dfrac{2\pi}{3}$ is a local minimum, so $x_2 = \dfrac{2\pi}{3}$ is a point of stable equilibrium.

10.61. Model: Assume $U = c/x$. Use conservation of energy.
Solve:
(a)

$$U_i + K_i = U_f + K_f$$

K_i is zero.

$$K_f = U_i - U_f$$
$$\frac{1}{2}mv^2 = \frac{c}{x_i} - \frac{c}{x_f}$$

$x_f = 1$

$$v^2 = \frac{2c}{m}\left(\frac{1}{x_i} - 1\right) = \frac{2c}{m}\frac{1}{x_i} - \frac{2c}{m}$$

This suggests that a graph of v^2 vs. $\dfrac{1}{x_i}$ would be a straight line with a slope of $2c/m$ and intercept of $-2c/m$.

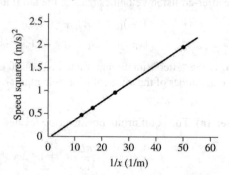

Speed Squared vs. Inverse Distance
$y = 0.04x - 0.0397, R^2 = 1$

x_i (m)	v_f (m/s)	$1/x_i$	v^2 (m/s)^2
0.02	1.4	50	1.96
0.04	0.98	25	0.9604
0.06	0.79	16.6666667	0.6241
0.08	0.68	12.5	0.4624

The spreadsheet graph shows the linear fit of v^2 vs. $\dfrac{1}{x_i}$ is extraordinary and that the slope and the intercept have the same magnitude, as our theoretical considerations predicted.

(b) We can find the value of c from either the slope or the intercept. The spreadsheet reports the slope as $0.040 \, \text{m}^3/\text{s}^2$.

$$c = -\frac{m}{2}(\text{slope}) = -\frac{0.050 \text{ kg}}{2}(0.040 \text{ m}^3/\text{s}^2) = 0.0010 \text{ J/m} = 1.0 \text{ mJ/m}$$

Assess: The units of the rise/run of the graph are m^3/s^2 and from the original statement of U the units of c should be J/m, and those are consistent with each other in part (b).

11

WORK

Exercises and Problems

Section 11.2 Work and Kinetic Energy

Section 11.3 Calculating and Using Work

11.3. Solve: (a) The length of $\vec{A}$ is $|\vec{A}| = A = \sqrt{\vec{A} \cdot \vec{A}} = \sqrt{(3)^2 + (4)^2} = \sqrt{25} = 5.$ The length of $\vec{B}$ is $B = \sqrt{(2)^2 + (-6)^2} = \sqrt{40} = 2\sqrt{10}.$ Using the answer $\vec{A} \cdot \vec{B} = -18$ from Ex 11.1(a),

$$\vec{A} \cdot \vec{B} = AB \cos\alpha$$
$$-18 = (5)(2\sqrt{10})\cos\alpha$$
$$\alpha = \cos^{-1}(-18/\sqrt{40}) = 125°$$

(b) The length of $\vec{A}$ is $|\vec{A}| = A = \sqrt{\vec{A} \cdot \vec{A}} = \sqrt{(3)^2 + (-2)^2} = \sqrt{13}.$ The length of $\vec{B}$ is $B = \sqrt{(6)^2 + (4)^2} = \sqrt{52} = 2\sqrt{13}.$ Using the answer $\vec{A} \cdot \vec{B} = 10$ from Ex 11.1(b),

$$\vec{A} \cdot \vec{B} = AB \cos\alpha$$
$$10 = (\sqrt{13})(2\sqrt{13})\cos\alpha$$
$$\alpha = \cos^{-1}(10/26) = 67°$$

11.5. Visualize: Please refer to Figure EX11.5.
Solve: (a) $\vec{A} \cdot \vec{B} = AB \cos\alpha = (5)(3)\cos 40° = 11.$

(b) $\vec{C} \cdot \vec{D} = CD \cos\alpha = (2)(3)\cos 140° = -4.6.$

(c) $\vec{E} \cdot \vec{F} = EF \cos\alpha = (3)(4)\cos 90° = 0.$

11.7. Solve: (a) $W = \vec{F} \cdot \Delta\vec{r} = (-3.0\hat{i} + 6.0\hat{j}) \cdot (2.0\hat{i})\,\text{N m} = \left(-6.0\hat{i} \cdot \hat{i} + 12\overset{=0}{\hat{j} \cdot \hat{i}} \right)\text{J} = -6.0\text{ J}.$

(b) $W = \vec{F} \cdot \Delta\vec{r} = (-3.0\hat{i} + 6.0\hat{j}) \cdot (2.0\hat{j})\,\text{N m} = \left(-6.0\overset{=0}{\hat{i} \cdot \hat{j}} + 12\hat{j} \cdot \hat{j} \right)\text{J} = 12\text{ J}.$

11.11. Model: Model the piano as a particle and use $W = \vec{F} \cdot \Delta\vec{r}$, where W is the work done by the force $\vec{F}$ through the displacement $\Delta\vec{r}$.

Visualize:

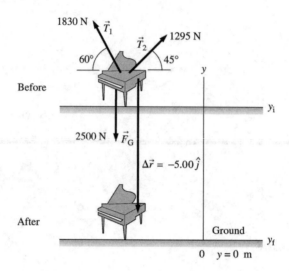

Solve: For the force $\vec{F}_G$:

$$W = \vec{F} \cdot \Delta \vec{r} = \vec{F}_G \cdot \Delta \vec{r} = (F_g) \cdot (\Delta r) \cos(0°) = (255 \text{ kg})(9.81 \text{ m/s}^2)(5.00 \text{ m})(1.00) = 1.25 \times 10^4 \text{ J}$$

For the tension $\vec{T}_1$:

$$W = \vec{T}_1 \cdot \Delta \vec{r} = (T_1)(\Delta r) \cos(150°) = (1830 \text{ N})(5.00 \text{ m})(-0.8660) = -7.92 \times 10^3 \text{ J}$$

For the tension $\vec{T}_2$:

$$W = \vec{T}_2 \cdot \Delta \vec{r} = (T_2)(\Delta r) \cos(135°) = (1295 \text{ N})(5.00 \text{ m})(-0.7071) = -4.58 \times 10^3 \text{ J}$$

Assess: Note that the displacement $\Delta \vec{r}$ in all the above cases is directed downwards along $-\hat{j}$.

11.13. Model: Model the 2.0 kg object as a particle, and use the work–kinetic-energy theorem.
Visualize: Please refer to Figure EX11.13. For each of the five intervals the velocity-versus-time graph gives the initial and final velocities. The mass of the object is 2.0 kg.
Solve: According to the work–kinetic-energy theorem:

$$W = \Delta K = \tfrac{1}{2}mv_f^2 = \tfrac{1}{2}mv_i^2 = \tfrac{1}{2}m(v_f^2 - v_i^2)$$

Interval AB: $v_i = 2 \text{ m/s}, v_f = -2 \text{ m/s} \Rightarrow W = \tfrac{1}{2}(2.0 \text{ kg})\left[(-2 \text{ m/s})^2 - (2 \text{ m/s})^2\right] = 0 \text{ J}$

Interval BC: $v_i = -2 \text{ m/s}, v_f = -2 \text{ m/s} \Rightarrow W = \tfrac{1}{2}(2.0 \text{ kg})\left[(-2 \text{ m/s})^2 - (-2 \text{ m/s})\right]^2 = 0 \text{ J}$

Interval CD: $v_i = -2 \text{ m/s}, v_f = 0 \text{ m/s} \Rightarrow W = \tfrac{1}{2}(2.0 \text{ kg})\left[(0 \text{ m/s})^2 - (-2 \text{ m/s})^2\right] = -4 \text{ J}$

Interval DE: $v_i = 0 \text{ m/s}, v_f = 2 \text{ m/s} \Rightarrow W = \tfrac{1}{2}(2.0 \text{ kg})\left[(2 \text{ m/s})^2 - (0 \text{ m/s})^2\right] = +4 \text{ J}$

Assess: The work done is zero in intervals AB and BC. In the interval CD + DE the total work done is zero. It is not whether v is positive or negative that counts because $K \propto v^2$. What is important is the magnitude of v and how v changes.

Section 11.4 The Work Done by a Variable Force

11.15. Model: Use the work–kinetic-energy theorem to find velocities.
Visualize: Please refer to Figure EX11.15.

Solve: The work–kinetic-energy theorem is

$$\Delta K = \tfrac{1}{2}mv_f^2 - \tfrac{1}{2}mv_i^2 = W = \int_{x_i}^{x_f} F_x \, dx = \text{ area under the force curve from } x_i \text{ to } x_f$$

$$\tfrac{1}{2}mv_f^2 - \tfrac{1}{2}(0.500 \text{ kg})(2.0 \text{ m/s})^2 = \tfrac{1}{2}mv_f^2 - 1.0 \text{ J} = \int_{0 \text{ m}}^{x_f} F_x \, dx = \tfrac{5}{2}x^2 \text{ Nm}$$

$$v_f = \sqrt{\frac{5x^2 \text{ Nm}}{0.500 \text{ kg}} + 4.0 \text{ m}^2/\text{s}^2}$$

At $x = 1$ m: $\Rightarrow$ $v_f = 3.7$ m/s

At $x = 2$ m: $\Rightarrow$ $v_f = 6.6$ m/s

At $x = 3$ m: $\Rightarrow$ $v_f = 9.7$ m/s

Section 11.5 Work and Potential Energy

Section 11.6 Finding Force from Potential Energy

11.21. Model: Use the negative derivative of the potential energy to determine the force acting on a particle.
Solve: **(b)** The x-component of the force is

$$F_x = -\frac{dU}{dx} = -\frac{d}{dx}\left(\frac{10}{x}\right) = \frac{10}{x^2}$$

$$F(x = 2 \text{ m}) = \frac{10}{x^2}\bigg|_{x=2 \text{ m}} = 2.5 \text{ N}, \; F(x = 5 \text{ m}) = \frac{10}{x^2}\bigg|_{x=5 \text{ m}} = 0.40 \text{ N}, \; F(= 8 \text{ m}) = \frac{10}{x^2}\bigg|_{x=8 \text{ m}} = 0.16 \text{ N}$$

Section 11.7 Thermal Energy

11.23. Visualize: One mole of helium atoms in the gas phase contains $N_A = 6.02 \times 10^{23}$ atoms.
Solve: If each atom moves with the same speed v, the microscopic total kinetic energy will be

$$K_{micro} = N_A\left(\frac{1}{2}mv^2\right) = 3700 \text{ J} \quad \Rightarrow \quad v = \sqrt{\frac{2K_{micro}}{mN_A}} = \sqrt{\frac{2(3700 \text{ J})}{(6.68 \times 10^{-27} \text{ kg})(6.02 \times 10^{23})}} = 1360 \text{ m/s}$$

Section 11.8 Conservation of Energy

11.27. Solve: Please refer to Figure EX11.27. The energy conservation equation yields
$$K_i + U_i + W_{ext} = K_f + U_f + \Delta E_{th} \quad \Rightarrow \quad 4 \text{ J} + 1 \text{ J} + W_{ext} = 1 \text{ J} + 2 \text{ J} + 1 \text{ J} \quad \rightarrow \quad W_{ext} - -1 \text{ J}$$
Thus, the work done to the environment is -1 J. In other words, 1 J of energy is transferred from the system into the environment. This is shown in the energy bar chart.

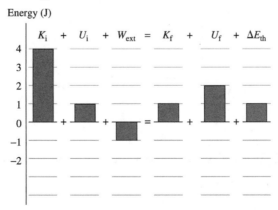

Section 11.9 Power

11.29. Model: Model the elevator as a particle, and apply the conservation of energy.
Solve: The tension in the cable does work on the elevator to lift it. Because the cable is pulled by the motor, we say that the motor does the work of lifting the elevator.
(a) The energy conservation equation is $K_i + U_i + W_{ext} = K_f + U_f + \Delta E_{th}$. Using $K_i = 0$ J, $K_f = 0$ J, and $\Delta E_{th} = 0$ J gives

$$W_{ext} = (U_f - U_i) = mg(y_f - y_i) = (1000 \text{ kg})(9.8 \text{ m/s}^2)(100 \text{ m}) = 9.80 \times 10^5 \text{ J}$$

(b) The power required to give the elevator this much energy in a time of 50 s is

$$P = \frac{W_{ext}}{\Delta t} = \frac{9.80 \times 10^5 \text{ J}}{50 \text{ s}} = 1.96 \times 10^4 \text{ W}$$

Assess: Since 1 horsepower (hp) is 746 W, the power of the motor is 26 hp. This is a reasonable amount of power to lift a mass of 1000 kg to a height of 100 m in 50 s.

11.31. Solve: The power of the solar collector is the solar energy collected divided by time. The intensity of the solar energy striking the earth is the power divided by area. We have

$$P = \frac{\Delta E}{\Delta t} = \frac{150 \times 10^6 \text{ J}}{3600 \text{ s}} = 41,667 \text{ W and intensity} = 1000 \text{ W/m}^2$$

$$\text{Area of solar collector} = \frac{41,667 \text{ W}}{1000 \text{ W/m}^2} = 42 \text{ m}^2$$

11.33. Solve: Using the conversion $746 \text{ W} = 1 \text{ hp}$, we have a power of 1492 J/s. This means $W = Pt = (1492 \text{ J/s})(1 \text{ h}) = 5.3712 \times 10^6 \text{ J}$ is the total work done by the electric motor in one hour. Furthermore,

$$W_{motor} = -W_g = U_{gf} - U_{gi} = mg(y_f - y_i) = mg(10 \text{ m})$$

$$m = \frac{W_{motor}}{g(10 \text{ m})} = \frac{5.3712 \times 10^6 \text{ J}}{(9.8 \text{ m/s}^2)(10 \text{ m})} = 5.481 \times 10^4 \text{ kg} = 5.481 \times 10^4 \text{ kg} \times \frac{1 \text{ liter}}{1 \text{ kg}} = 5.5 \times 10^4 \text{ liters}$$

11.39. Model: Use the relationship between force and potential energy and the work–kinetic-energy theorem.
Visualize: Please refer to Figure P11.39. We will find the slope in the following x regions: $0 \text{ cm} < x < 1 \text{ cm}$, $1 < x < 3 \text{ cm}$, $3 < x < 5 \text{ cm}$, $5 < x < 7 \text{ cm}$, and $7 < x < 8 \text{ cm}$.
Solve: (a) F_x is the negative slope of the U-versus-x graph, for example, for $0 \text{ m} < x < 2 \text{ m}$

$$\frac{dU}{dx} = \frac{-4 \text{ J}}{0.01 \text{ m}} = -400 \text{ N} \quad \Rightarrow \quad F_x = +400 \text{ N}$$

Calculating the values of F_x in this way, we can draw the force-versus-position graph as shown below.

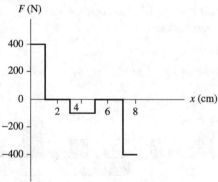

(b) Since $W = \int_{x_i}^{x_f} F_x \, dx =$ area of the F_x-versus-x graph between x_i and x_f, the work done by the force as the particle moves from $x_i = 2 \text{ cm}$ to $x_f = 6 \text{ cm}$ is -2 J.

(c) The conservation of energy equation is $K_f + U_f = K_i + U_i$. We can see from the graph that $U_i = 0$ J and $U_f = 2$ J in moving from $x = 2$ cm to $x = 6$ cm. The final speed is $v_f = 10$ m/s, so

$$2 \text{ J} + \tfrac{1}{2}(0.010 \text{ kg})(10.0 \text{ m/s})^2 = 0 \text{ J} + \tfrac{1}{2}(0.010 \text{ kg})v_i^2 \quad \Rightarrow \quad v_i = 22 \text{ m/s}$$

11.41. Model: Use $a_x = dv_x/dt$, $x = \int v_x\, dt$, $K = \tfrac{1}{2}mv_x^2$, and $F = ma_x$.

Visualize: Please refer to Figure P11.41. We know $a_x = $ slope of the v_x-versus-t graph and $x = $ area under the v_x-versus-x graph between 0 and x.

Solve: Using the above definitions and methodology, we can generate the following table:

t(s)	a_x (m/s2)	x(m)	K(J)	F(N)
0	10	0	0	5
0.5	10	1.25	6.25	5
1.0	10	5	25	5
1.5	10	11.25	56.25	5
2.0	+10 or −10	20	100	−5 or +5
2.5	10	28.75	56.25	5
3.0	−10 or 0	35	25	−5 or 0
3.5	0	40	25	0
4.0	0	45	25	0

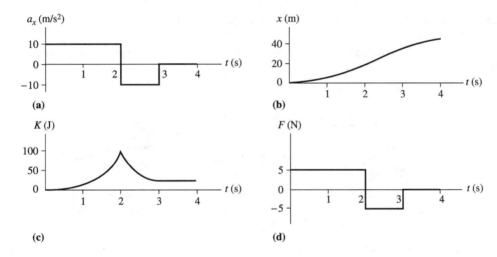

(a) (b)

(c) (d)

(e) Let J_1 be the impulse from $t = 0$ s to $t = 2$ s and J_2 be the impulse from $t = 2$ s to $t = 4$ s. We have

$$J_1 = \int_{0s}^{2s} F_x\, dt = (5 \text{ N})(2 \text{ s}) = 10 \text{ N} \cdot \text{s} \quad \text{and} \quad J_2 = \int_{0s}^{4s} F_x\, dt = (-5 \text{ N})(1 \text{ s}) = -5 \text{ N} \cdot \text{s}$$

(f) $J = \Delta p = mv_f - mv_i \quad \Rightarrow \quad v_f = v_i + J/m$

At $t = 2$ s, $v_x = 0$ m/s + (10 N·s)/(0.5 kg) = 0 m/s + 20 m/s = 20 m/s

At $t = 4$ s, $v_x = 20$ m/s + (−5 N·s)/(0.5 kg) = 20 m/s − 10 m/s = 10 m/s

The v_x-versus-t graph also gives $v_x = 20$ m/s at $t = 2$ s and $v_x = 10$ m/s at $t = 4$ s.

(g)

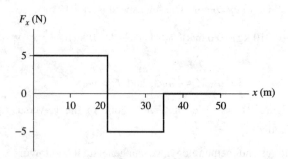

(h) From $t = 0$ s to $t = 2$ s, $W = \int F_x\, dx = (5\ \text{N})(20\ \text{m}) = 100\ \text{J}$

From $t = 2$ s to $t = 4$ s, $W = \int F_x\, dx = (-5\ \text{N})(15\ \text{m}) = -75\ \text{J}$

(i) At $t = 0$ s, $v_x = 0$ m/s so the work–kinetic-energy theorem for calculating v_x at $t = 2$ s is

$$W = \Delta K = \frac{1}{2}mv_{\text{f}}^2 - \frac{1}{2}mv_{\text{i}}^2 \quad \Rightarrow \quad 100\ \text{J} = \frac{1}{2}(0.5\ \text{kg})v_x^2 - \frac{1}{2}(0.5\ \text{kg})(0\ \text{m/s})^2 \quad \Rightarrow \quad v_x = 20\ \text{m/s}$$

To calculate v_x at $t = 4$ s, we use v_x at $t = 2$ s as the initial velocity:

$$-75\ \text{J} = \frac{1}{2}(0.5\ \text{kg})v_x^2 - \frac{1}{2}(0.5\ \text{kg})(20\ \text{m/s})^2 \quad \Rightarrow \quad v_x = 10\ \text{m/s}$$

Both of these values agree with the values on the velocity graph.

11.45. Model: Model Sam strapped with skis as a particle, and apply the law of conservation of energy.
Visualize:

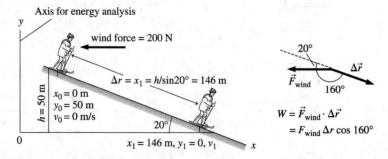

Solve: (a) The conservation of energy equation is

$$K_1 + U_{\text{g1}} + \Delta E_{\text{th}} = K_0 + U_{\text{g0}} + W_{\text{ext}}$$

The snow is frictionless, so $\Delta E_{\text{th}} = 0$ J. However, the wind is an external force doing work on Sam as he moves down the hill. Thus,

$$W_{\text{ext}} = W_{\text{wind}} = (K_1 + U_{\text{g1}}) - (K_0 + U_{\text{g0}})$$

$$= \left(\tfrac{1}{2}mv_1^2 + mgy_1\right) - \left(\tfrac{1}{2}mv_0^2 + mgy_0\right) = \left(\tfrac{1}{2}mv_1^2 + 0\ \text{J}\right) - (0\ \text{J} + mgy_0) = \tfrac{1}{2}mv_1^2 - mgy_0$$

$$v_1 = \sqrt{2gy_0 + \frac{2W_{\text{wind}}}{m}}$$

We compute the work done by the wind as follows:

$$W_{\text{wind}} = \vec{F}_{\text{wind}} \cdot \Delta \vec{r} = F_{\text{wind}} \Delta r \cos(160°) = (200\ \text{N})(146\ \text{m})\cos(160°) = -27,400\ \text{J}$$

where we have used $\Delta r = h/\sin(20°) = 146$ m. Now we can compute

$$v_1 = \sqrt{2(9.8\ \text{m/s}^2)(50\ \text{m}) + \frac{2(-27,400\ \text{J})}{75\ \text{kg}}} = 16\ \text{m/s}$$

Assess: We used a vertical y-axis for energy analysis, rather than a tilted coordinate system, because U_g is determined by its vertical position.

11.47. Model: Assume an ideal spring that obeys Hooke's law. Model the box as a particle and use the model of kinetic friction.
Visualize:

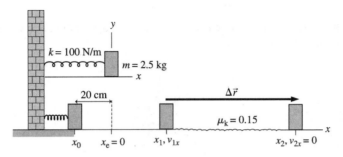

Solve: When the horizontal surface is frictionless, conservation of energy means
$$\tfrac{1}{2}k(x_0 - x_e)^2 = \tfrac{1}{2}mv_{1x}^2 = K_1 \quad \Rightarrow \quad K_1 = \tfrac{1}{2}(100 \text{ N/m})(0.20 \text{ m} - 0 \text{ m})^2 = 2.0 \text{ J}$$
That is, the box is launched with 2.0 J of kinetic energy. It will lose 2.0 J of kinetic energy on the rough surface. The net force on the box is $\vec{F}_{net} = -\vec{f}_k = -\mu_k mg \hat{i}$. The work–kinetic-energy theorem is
$$W_{net} = \vec{F}_{net} \cdot \Delta \vec{r} = K_2 - K_1 = 0 \text{ J} - 2.0 \text{ J} = -2.0 \text{ J}$$
$$(-\mu_k mg)(x_2 - x_1) = -2.0 \text{ J}$$
$$(x_2 - x_1) = \frac{2.0 \text{ J}}{\mu_k mg} = \frac{2.0 \text{ J}}{(0.15)(2.5 \text{ kg})(9.8 \text{ m/s}^2)} = 0.54 \text{ m}$$

Assess: Because the force of friction transforms kinetic energy into thermal energy, energy is transferred out of the box into the environment. In response, the box slows down and comes to rest.

11.49. Model: Identify the truck and the loose gravel as the system. We need the gravel inside the system because friction increases the temperature of the truck and the gravel. We will also use the model of kinetic friction and the conservation of energy equation.
Visualize:

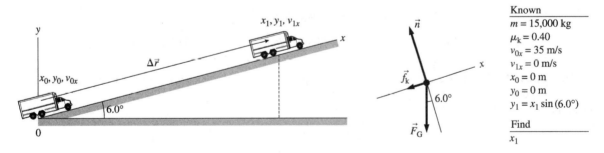

Known
$m = 15,000$ kg
$\mu_k = 0.40$
$v_{0x} = 35$ m/s
$v_{1x} = 0$ m/s
$x_0 = 0$ m
$y_0 = 0$ m
$y_1 = x_1 \sin(6.0°)$

Find
x_1

We place the origin of our coordinate system at the base of the ramp in such a way that the x-axis is along the ramp and the y-axis is vertical so that we can calculate potential energy. The free-body diagram of forces on the truck is shown.
Solve: The conservation of energy equation is $K_1 + U_{g1} + \Delta E_{th} = K_0 + U_{g0} + W_{ext}$. In the present case, $W_{ext} = 0$ J, $v_{1x} = 0$ m/s, $U_{g0} = 0$ J, $v_{0x} = 35$ m/s. The thermal energy created by friction is
$$\Delta E_{th} = f_k(x_1 - x_0) = (\mu_k n)(x_1 - x_0) = \mu_k mg \cos(6.0°)(x_1 - x_0)$$
$$= (0.40)(15,000 \text{ kg})(9.8 \text{ m/s}^2)\cos(6.0°)(x_1 - x_0) = (58,478 \text{ J/m})(x_1 - x_0)$$

Thus, the energy conservation equation simplifies to

$$0\ \text{J} + mgy_1 + (58{,}478\ \text{J/m})(x_1 - x_0) = \tfrac{1}{2}mv_{0x}^2 + 0\ \text{J} + 0\ \text{J}$$

$$(15{,}000\ \text{kg})(9.8\ \text{m/s}^2)(x_1 - x_0)\sin(6.0°) + (58{,}478\ \text{J/m})(x_1 - x_0) = \tfrac{1}{2}(15{,}000\ \text{kg})(35\ \text{m/s})^2$$

$$(x_1 - x_0) = 124\ \text{m} = 0.12\ \text{km}$$

Assess: A length of 124 m at a slope of 6° seems reasonable.

11.53. Model: Model the water skier as a particle, apply the law of conservation of mechanical energy, and use the constant-acceleration kinematic equations.
Visualize:

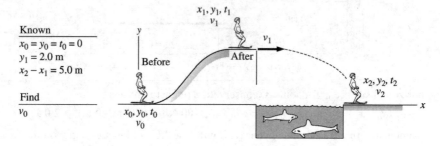

We placed the origin of the coordinate system at the base of the frictionless ramp.
Solve: We'll start by finding the smallest speed v_1 at the top of the ramp that allows her to clear the shark tank. From the vertical motion for jumping the shark tank,

$$y_2 = y_1 + v_{1y}\Delta t + \tfrac{1}{2}a_y \Delta t^2$$

$$0\ \text{m} = 2.0\ \text{m} + 0\ \text{m} + \tfrac{1}{2}(9.8\ \text{m/s}^2)\Delta t^2 \quad \Rightarrow \quad \Delta t = 0.639\ \text{s}$$

From the horizontal motion,

$$x_2 = x_1 + v_{1x}\Delta t + \tfrac{1}{2}a_x \Delta t^2$$

$$(x_1 + 5.0\ \text{m}) = x_1 + v_1\Delta t + 0\ \text{m} \quad \Rightarrow \quad v_1 = \frac{5.0\ \text{m}}{0.639\ \text{s}} = 7.825\ \text{m/s}$$

Having found the v_1 that will take the skier to the other side of the tank, we now use the energy equation to find the minimum speed v_0. We have

$$K_1 + U_{g1} = K_0 + U_{g0} \quad \Rightarrow \quad \tfrac{1}{2}mv_1^2 + mgy_1 = \tfrac{1}{2}mv_0^2 + mgy_0$$

$$v_0 = \sqrt{v_1^2 + 2g(y_1 - y_0)} = \sqrt{(7.825\ \text{m/s})^2 + 2(9.8\ \text{m/s}^2)(2.0\ \text{m})} = 10\ \text{m/s}$$

11.55. Model: Model the ice cube as a particle, the spring as an ideal that obeys Hooke's law, and the law of conservation of energy.
Visualize:

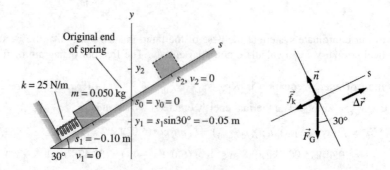

Solve: (a) The normal force does no work and the slope is frictionless, so mechanical energy is conserved. We've drawn two separate axes: a vertical y-axis to measure potential energy and a tilted s-axis to measure distance along the slope. Both have the same origin which is at the point where the spring is not compressed. Thus, the two axes are related by $y = s \sin \theta$. Also, this choice of origin makes the elastic potential energy simply $U_s = \frac{1}{2} k (s - s_0)^2 = \frac{1}{2} k s^2$.

Because energy is conserved, we can relate the initial point—with the spring compressed—to the final point where the ice cube is at maximum height. We do *not* need to find the speed with which it leaves the spring. We have
$$K_2 + U_{g2} + U_{s2} = K_1 + U_{g1} + U_{s1}$$
$$\tfrac{1}{2} m v_2^2 + m g y_2 + \tfrac{1}{2} k s_0^2 = \tfrac{1}{2} m v_1^2 + m g y_1 + \tfrac{1}{2} k s_1^2$$

It is important to note that at the final point, when the ice cube is at y_2, the end of the spring is only at s_0. The spring does *not* stretch to s_2, so U_{s2} is *not* $\frac{1}{2} k s_2^2$. Three of the terms are zero, leaving
$$m g y_2 = + m g y_1 + \tfrac{1}{2} k s_1^2 \quad \Rightarrow \quad y_2 - y_1 = \Delta y = \text{height gained} = \frac{k s_1^2}{2 m g} = 0.255 \text{ m} = 25.5 \text{ cm}$$

The distance traveled is $\Delta s = \Delta y / \sin(30°) = 0.51$ m.

(b) Using the energy equation and the expression for thermal energy:
$$K_2 + U_{g2} + U_{s2} + \Delta E_{th} = K_1 + U_{g1} + U_{s1} + W_{ext}, \quad \Delta E_{th} = f_k \Delta s = \mu_k n \Delta s$$

From the free-body diagram,
$$(\vec{F}_{net})_y = 0 \text{ N} = n - m g \cos(30°) \quad \Rightarrow \quad n = m g \cos(30°)$$

Now, having found $\Delta E_{th} = \mu_k m g \cos(30°) \Delta s$, the energy equation can be written
$$0 \text{ J} + m g y_2 + 0 \text{ J} + \mu_k m g \cos(30°) \Delta s = 0 \text{ J} + m g y_1 + \tfrac{1}{2} k s_1^2 + 0 \text{ J}$$
$$m g (y_2 - y_1) - \tfrac{1}{2} k s_1^2 + \mu_k m g \cos(30°) \Delta s = 0$$

Using $\Delta y = \Delta s \sin(30°)$, the above equation simplifies to
$$m g \Delta s \sin(30°) + \mu_k m g \cos(30°) \Delta s = \tfrac{1}{2} k s_1^2 \quad \Rightarrow \quad \Delta s = \frac{k s_1^2}{2 m g [\sin(30°) + \mu_k \cos(30°)]} = 0.38 \text{ m}$$

11.57. Model: Assume an ideal spring, so Hooke's law is obeyed. Treat the physics student as a particle and apply the law of conservation of energy. Our system is comprised of the spring, the student, and the ground. We also use the model of kinetic friction.

Visualize: We place the origin of the coordinate system on the ground directly below the end of the compressed spring that is in contact with the student.

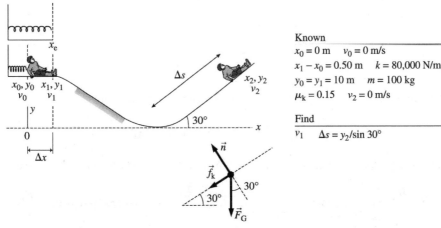

Known
$x_0 = 0$ m $v_0 = 0$ m/s
$x_1 - x_0 = 0.50$ m $k = 80,000$ N/m
$y_0 = y_1 = 10$ m $m = 100$ kg
$\mu_k = 0.15$ $v_2 = 0$ m/s

Find
v_1 $\Delta s = y_2 / \sin 30°$

Solve: (a) The energy conservation equation gives
$$K_1 + U_{g1} + U_{s1} + \Delta E_{th} = K_0 + U_{g0} + U_{s0} + W_{ext}$$
$$\tfrac{1}{2} m v_1^2 + m g y_1 + \tfrac{1}{2} k (x_1 - x_e)^2 + 0 \text{ J} = \tfrac{1}{2} m v_0^2 + m g y_0 + \tfrac{1}{2} k (x_1 - x_0)^2 + 0 \text{ J}$$

Since $y_1 = y_0 = 10$ m, $x_1 = x_e$, $v_0 = 0$ m/s, $k = 80,000$ N/m, $m = 100$ kg, and $(x_1 - x_0) = 0.5$ m,

$$\tfrac{1}{2}mv_1^2 = \tfrac{1}{2}k(x_1 - x_0)^2 \quad \Rightarrow \quad v_1 = \sqrt{\frac{k}{m}}(x_1 - x_0) = \sqrt{\frac{80,000 \text{ N/}m}{100 \text{ kg}}}(0.50 \text{ m}) = 14 \text{ m/s}$$

(b) Friction creates thermal energy. Applying the conservation of energy equation once again:

$$K_2 + U_{g2} + U_{s2} + \Delta E_{th} = K_0 + U_{g0} + U_{s0} + W_{ext}$$

$$\tfrac{1}{2}mv_2^2 + mgy_2 + 0 \text{ J} + f_k\Delta s = 0 \text{ J} + mgy_0 + \tfrac{1}{2}k(x_1 - x_0)^2 + 0 \text{ J}$$

With $v_2 = 0$ m/s and $y_2 = \Delta s \sin(30°)$, the above equation is simplified to

$$mg\Delta s \sin(30°) + \mu_k n\Delta s = mgy_0 + \tfrac{1}{2}k(x_1 - x_0)^2$$

From the free-body diagram for the physics student, we see that $n = F_G \cos(30°) = mg \cos(30°)$. Thus, the conservation of energy equation gives

$$\Delta s[mg \sin(30°) + \mu_k mg \cos(30°)] = mgy_0 + \tfrac{1}{2}k(x_1 - x_0)^2$$

Using $m = 100$ kg, $k = 80,000$ N/m, $(x_1 - x_0) = 0.50$ m, $y_0 = 10$ m, and $\mu_k = 0.15$, we get

$$\Delta s = \frac{mgy_0 + \tfrac{1}{2}k(x_1 - x_0)^2}{mg[\sin(30°) + \mu_k \cos(30°)]} = 32 \text{ m}$$

Assess: $y_2 = \Delta s \sin(30°) = 16$ m, which is greater than $y_0 = 10$ m. The higher value is due to the transformation of the spring energy into gravitational potential energy.

11.61. Solve: (a) Because $\sin(cx)$ is dimensionless, F_0 must have units of force in newtons.
(b) The product cx is an angle because we are taking the sine of it. An angle is dimensionless. If x has units of m and the product cx is dimensionless, then c has to have units of m^{-1}.
(c) The force is a maximum when $\sin(cx) = 1$. This occurs when $cx = \pi/2$, or for $x_{max} = \pi/(2c)$.
(d) The graph is the first quarter of a sine curve.

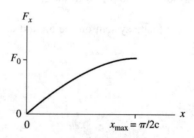

(e) We can find the velocity v_f at $x_f = x_{max}$ from the work–kinetic-energy theorem:

$$\Delta K = \tfrac{1}{2}mv_f^2 - \tfrac{1}{2}mv_i^2 = \tfrac{1}{2}mv_f^2 - \tfrac{1}{2}mv_0^2 = W \quad \Rightarrow \quad v_f = \sqrt{v_0^2 + \frac{2W}{m}}$$

This is a variable force. As the particle moves from $x_i = 0$ m to $x_f = x_{max} = \pi/(2c)$ the work done on it is

$$W = \int_{x_i}^{x_f} F(x)dx = F_0 \int_{x_i}^{x_f} \sin(cx)dx = -\frac{F_0}{c}\cos(cx)\Big|_0^{\pi/(2c)} = -\frac{F_0}{c}[\cos(\pi/2) - \cos(0)] = \frac{F_0}{c}$$

Thus, the particle's speed at $x_f = x_{max} = \pi/(2c)$ is $v_f = \sqrt{v_0^2 + 2F_0/(mc)}$.

11.65. Solve: (a) The change in the potential energy of 1.0 kg of water in falling 25 m is

$$\Delta U_g = -mgh = -(1.0 \text{ kg})(9.8 \text{ m/s}^2)(25 \text{ m}) = -245 \text{ J} \approx -0.25 \text{ kJ}$$

(b) The power required of the dam is

$$P = \frac{W}{t} = \frac{W}{1 \text{ s}} = 50 \times 10^6 \text{ Watts} \quad \Rightarrow \quad W = 50 \times 10^6 \text{ J}$$

That is, 50×10^6 J of energy is required per second for the dam. Out of the 245 J of lost potential energy, $(245 \text{ J})(0.80) = 196 \text{ J}$ is converted to electrical energy. Thus, the amount of water needed per second is $(50 \times 10^6 \text{ J})(1.0 \text{ kg}/196 \text{ J}) = 255{,}000 \text{ kg} \approx 2.6 \times 10^5 \text{ kg}$.

11.67. Model: Use the model of static friction, kinematic equations, and the definition of power.
Solve: (a) The rated power of the Porsche is 217 hp $= 161{,}882$ W and the gravitational force on the car is $(1480 \text{ kg})(9.8 \text{ m/s}^2) = 14{,}504$ N. The amount of that force on the drive wheels is $(14{,}504)(2/3) = 9670$ N. Because the static friction of the tires on road pushes the car forward,

$$F_{max} = f_{s,max} = \mu_s n = \mu_s mg = (1.00)(9670 \text{ N}) = ma_{max}$$

$$a_{max} = \frac{9670 \text{ N}}{1480 \text{ kg}} = 6.53 \text{ m/s}^2$$

(b) Only 70% of the power generated by the motor is applied at the wheels.

$$P = Fv_{max} \quad \Rightarrow \quad v_{max} = \frac{P}{F} = \frac{(0.70)(161{,}882 \text{ W})}{9670 \text{ N}} = 11.7 \text{ m/s}$$

(c) Using the kinematic equation, $v_{max} = v_0 + a_{max}(t_{min} - t_0)$ with $v_0 = 0$ m/s and $t_0 = 0$ s, we obtain

$$t_{min} = \frac{v_{max}}{a_{max}} = \frac{11.7 \text{ m/s}}{6.53 \text{ m/s}^2} = 1.79 \text{ s}$$

Assess: An acceleration time of 1.79 s for the Porsche to reach a speed of ≈ 26 mph from rest is reasonable.

ROTATION OF A RIGID BODY

Exercises and Problems

Section 12.1 Rotational Motion

12.1. Model: A spinning skater, whose arms are outstretched, is a rigid rotating body.
Visualize:

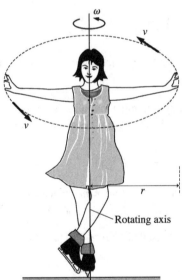

Solve: The speed $v = r\omega$, where $r = 140$ cm$/2 = 0.70$ m. Also, 180 rpm $= (180)2\pi/60$ rad/s $= 6\pi$ rad/s. Thus, $v = (0.70$ m$)(6\pi$ rad/s$) = 13.2$ m/s.
Assess: A speed of 13.2 m/s ≈ 26 mph for the hands is a little high, but reasonable.

Section 12.2 Rotation About the Center of Mass

12.5. Model: The earth and moon are particles.
Visualize:

Known
$x_E = 0$ m
$x_M = 13.84 \times 10^8$ m
$m_E = 5.98 \times 10^{24}$ kg
$m_M = 7.36 \times 10^{22}$ kg

Choosing $x_E = 0$ m sets the coordinate origin at the center of the earth so that the center of mass location is the distance from the center of the earth.

Solve:

$$x_{cm} = \frac{m_E x_E + m_M x_M}{m_E + m_M} = \frac{(5.98 \times 10^{24} \text{ kg})(0 \text{ m}) + (7.36 \times 10^{22} \text{ kg})(3.84 \times 10^8 \text{ m})}{5.98 \times 10^{24} \text{ kg} + 7.36 \times 10^{22} \text{ kg}}$$

$$= 4.67 \times 10^6 \text{ m} \approx 4.7 \times 10^6 \text{ m}$$

Assess: The center of mass of the earth-moon system is called the barycenter and is located beneath the surface of the earth. Even though $x_E = 0$ m the earth influences the center of mass location because m_E is in the denominator of the expression for x_{cm}.

Section 12.3 Rotational Energy

12.9. Model: The earth is a rigid, spherical rotating body.

Solve: The rotational kinetic energy of the earth is $K_{rot} = \frac{1}{2} I \omega^2$. The moment of inertia of a sphere about its diameter (see Table 12.2) is $I = \frac{2}{5} M_{earth} R^2$ and the angular velocity of the earth is

$$\omega = \frac{2\pi \text{ rad}}{24 \times 3600 \text{ s}} = 7.27 \times 10^{-5} \text{ rad/s}$$

Thus, the rotational kinetic energy is

$$K_{rot} = \frac{1}{2} \left(\frac{2}{5} M_{earth} R^2 \right) \omega^2$$

$$= \frac{1}{5} (5.98 \times 10^{24} \text{ kg})(6.37 \times 10^6 \text{ m})^2 (7.27 \times 10^{-5} \text{ rad/s})^2 = 2.57 \times 10^{29} \text{ J}$$

12.11. Model: The triangle is a rigid body rotating about an axis through the center.

Visualize: Please refer to Figure EX12.11. Each 200 g mass is a distance r away from the axis of rotation, where r is given by

$$\frac{0.20 \text{ m}}{r} = \cos 30° \Rightarrow r = \frac{0.20 \text{ m}}{\cos 30°} = 0.2309 \text{ m}$$

Solve: (a) The moment of inertia of the triangle is $I = 3 \times mr^2 = 3(0.200 \text{ kg})(0.2309 \text{ m})^2 = 0.032 \text{ kg m}^2$.

(b) The frequency of rotation is given as 5.0 revolutions per s or 10π rad/s. The rotational kinetic energy is

$$K_{rot.} = \frac{1}{2} I \omega^2 = \frac{1}{2} (0.0320 \text{ kg m}^2)(10.0\pi \text{ rad/s})^2 = 15.8 \text{ J} = 16 \text{ J}$$

Section 12.4 Calculating Moment of Inertia

12.13. Model: The moment of inertia of any object depends on the axis of rotation. In the present case, the rotation axis passes through mass A and is perpendicular to the page.

Solve: (a) $x_{cm} = \dfrac{\sum m_i x_i}{\sum m_i} = \dfrac{m_A x_A + m_B x_B + m_C x_C + m_D x_D}{m_A + m_B + m_C + m_D}$

$$= \frac{(100 \text{ g})(0 \text{ m}) + (200 \text{ g})(0 \text{ m}) + (200 \text{ g})(0.10 \text{ m}) + (200 \text{ g})(0.10 \text{ m})}{100 \text{ g} + 200 \text{ g} + 200 \text{ g} + 200 \text{ g}} = 0.057 \text{ m}$$

$$y_{cm} = \frac{m_A y_A + m_B y_B + m_C y_C + m_D y_D}{m_A + m_B + m_C + m_D}$$

$$= \frac{(100 \text{ g})(0 \text{ m}) + (200 \text{ g})(0.08 \text{ m}) + (200 \text{ g})(0.08 \text{ cm}) + (200 \text{ g})(0 \text{ m})}{700 \text{ g}} = 0.046 \text{ m}$$

(b) The distance from the axis to mass C is 12.81 cm. The moment of inertia through A and perpendicular to the page is

$$I_A = \sum_i m_i r_i^2 = m_A r_A^2 + m_B r_B^2 + m_C r_C^2 + m_D r_D^2$$

$$= (0.100 \text{ kg})(0 \text{ m})^2 + (0.200 \text{ kg})(0.08 \text{ m})^2 + (0.200 \text{ kg})(0.1281 \text{ m})^2 + (0.200 \text{ kg})(0.10 \text{ m})^2 = 0.0066 \text{ kg m}^2$$

12.17. Model: The CD is a disk of uniform density.
Solve: (a) The center of the CD is its center of mass. Using Table 12.2,

$$I_{cm} = \frac{1}{2} MR^2 = \frac{1}{2}(0.021 \text{ kg})(0.060 \text{ m})^2 = 3.8 \times 10^{-5} \text{ kg m}^2$$

(b) Using the parallel–axis theorem with $d = 0.060$ m,

$$I = I_{cm} + Md^2 = 3.8 \times 10^{-5} \text{ kg m}^2 + (0.021 \text{ kg})(0.060 \text{ m})^2 = 1.14 \times 10^{-4} \text{ kg m}^2$$

Section 12.5 Torque

12.19. Visualize:

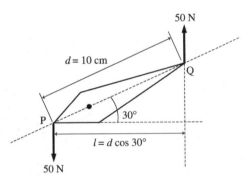

The two equal but opposite 50 N forces, one acting at point P and the other at point Q, make a couple that causes a net torque.
Solve: The distance between the lines of action is $l = d \cos 30°$. The net torque is given by
$$\tau = lF = (d \cos 30°)F = (0.10 \text{ m})(0.866)(50 \text{ N}) = 4.3 \text{ N m}$$

Section 12.6 Rotational Dynamics

Section 12.7 Rotation About a Fixed Axis

12.23. Solve: $\tau = I\alpha$ is the rotational analog of Newton's second law $F = ma$. We have
$\tau = (2.0 \text{ kg m}^2)(4.0 \text{ rad/s}^2) = 8.0 \text{ kg m}^2/\text{s}^2 = 8.0 \text{ N m}.$

12.25. Model: Two balls connected by a rigid, massless rod are a rigid body rotating about an axis through the center of mass. Assume that the size of the balls is small compared to 1 m.
Visualize:

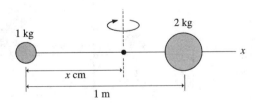

We placed the origin of the coordinate system on the 1.0 kg ball.

Solve: The center of mass and the moment of inertia are

$$x_{cm} = \frac{(1.0 \text{ kg})(0 \text{ m}) + (2.0 \text{ kg})(1.0 \text{ m})}{(1.0 \text{ kg} + 2.0 \text{ kg})} = 0.667 \text{ m} \quad \text{and} \quad y_{cm} = 0 \text{ m}$$

$$I_{about\ cm} = \sum m_i r_i^2 = (1.0 \text{ kg})(0.667 \text{ m})^2 + (2.0 \text{ kg})(0.333 \text{ m})^2 = 0.667 \text{ kg m}^2$$

We have $\omega_f = 0$ rad/s, $t_f - t_i = 5.0$ s, and $\omega_i = -20$ rpm $= -20(2\pi \text{ rad}/60 \text{ s}) = -\frac{2}{3}\pi \text{ rad/s}$, so $\omega_f = \omega_i + \alpha(t_f - t_i)$ becomes

$$0 \text{ rad/s} = \left(-\frac{2\pi}{3}\text{rad/s}\right) + \alpha(5.0 \text{ s}) \Rightarrow \alpha = \frac{2\pi}{15}\text{rad/s}^2$$

Having found I and α, we can now find the torque τ that will bring the balls to a halt in 5.0 s:

$$\tau = I_{about\ cm}\alpha = \left(\frac{2}{3}\text{kg m}^2\right)\left(\frac{2\pi}{15}\text{ rad/s}^2\right) = \frac{4\pi}{45} \text{ N m} = 0.28 \text{ N m}$$

The magnitude of the torque is 0.28 N m, applied in the counterclockwise direction.

12.27. Model: Model the rod as thin enough to use $I = \frac{1}{3}ML^2$.

Visualize:

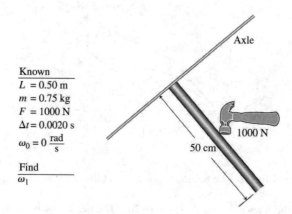

Known
L = 0.50 m
m = 0.75 kg
F = 1000 N
Δt = 0.0020 s
$\omega_0 = 0 \frac{\text{rad}}{\text{s}}$

Find
ω_1

Solve: From Newton's second law we have $\Delta L = \tau \Delta t$. Combine with $\Delta L = I\Delta\omega$ and then solve for ω_1.

$$\tau\Delta t = I\Delta\omega$$

With $\omega_0 = 0$ we have $\Delta\omega = \omega_1 - \omega_0 = \omega_1$. Also $\tau = rF$ where $r = 0.25$ m because the rod is hit perpendicular to it.

$$\omega_1 = \Delta\omega = \frac{\tau\Delta t}{I} = \frac{rF\Delta t}{\frac{1}{3}ML^2} = \frac{(0.25 \text{ m})(1000 \text{ N})(0.0020 \text{ s})}{\frac{1}{3}(0.75 \text{ kg})(0.50 \text{ m})^2} = 8.0 \text{ rad/s}$$

Assess: The units check out, and 8.0 rad/s seems like a reasonable answer.

Section 12.8 Static Equilibrium

12.31. Model: The see-saw is a rigid body. The cats and bowl are particles.
Visualize:

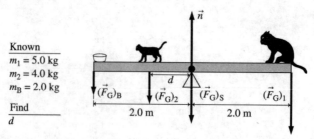

Known
$m_1 = 5.0$ kg
$m_2 = 4.0$ kg
$m_B = 2.0$ kg

Find
d

Solve: The see-saw is in rotational equilibrium. Calculate the net torque about the pivot point.

$$\tau_{\text{net}} = 0 = (F_G)_1(2.0 \text{ m}) - (F_G)_2(d) - (F_G)_B(2.0 \text{ m})$$

$$m_2gd = m_1g(2.0 \text{ m}) - m_Bg(2.0 \text{ m})$$

$$d = \frac{(m_1 - m_B)(2.0 \text{ m})}{m_2} = \frac{(5.0 \text{ kg} - 2.0 \text{ kg})(2.0 \text{ m})}{4.0 \text{ kg}} = 1.5 \text{ m}$$

Assess: The smaller cat is close but not all the way to the end by the bowl, which makes sense since the combined mass of the smaller cat and bowl of tuna is greater than the mass of the larger cat.

Section 12.9 Rolling Motion

12.35. Model: The mechanical energy of both the hoop (h) and the sphere (s) is conserved. The initial gravitational potential energy is transformed into kinetic energy as the objects roll down the slope. The kinetic energy is a combination of translational and rotational kinetic energy. We also assume no slipping of the hoop or of the sphere.
Visualize:

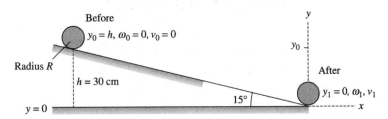

The zero of gravitational potential energy is chosen at the bottom of the slope.
Solve: The energy conservation equation for the sphere or hoop $K_f + U_{gf} = K_i + U_{gi}$ is

$$\frac{1}{2}I(\omega_1)^2 + \frac{1}{2}m(v_1)^2 + mgy_1 = \frac{1}{2}I(\omega_0)^2 + \frac{1}{2}m(v_0)^2 + mgy_0$$

For the sphere, this becomes

$$\frac{1}{2}\left(\frac{2}{5}mR^2\right)\frac{(v_1)_s^2}{R^2} + \frac{1}{2}m(v_1)_s^2 + 0 \text{ J} = 0 \text{ J} + 0 \text{ J} + mgh_s$$

$$\Rightarrow \frac{7}{10}(v_1)_s^2 = gh \Rightarrow (v_1)_s = \sqrt{10gh/7} = \sqrt{10(9.8 \text{ m/s}^2)(0.30 \text{ m})/7} = 2.05 \text{ m/s}$$

For the hoop, this becomes

$$\frac{1}{2}(mR^2)\frac{(v_1)_h^2}{R^2} + \frac{1}{2}m(v_1)_h^2 + 0 \text{ J} = 0 \text{ J} + 0 \text{ J} + mgh_{\text{hoop}}$$

$$\Rightarrow h_{\text{hoop}} = \frac{(v_1)_h^2}{g}$$

For the hoop to have the same velocity as that of the sphere,

$$h_{\text{hoop}} = \frac{(v_1)_s^2}{g} = \frac{(2.05 \text{ m/s})^2}{9.8 \text{ m/s}^2} = 42.9 \text{ cm}$$

The hoop should be released from a height of 43 cm.

Section 12.10 The Vector Description of Rotational Motion

12.39. Solve: (a) $\hat{i} \times (\hat{i} \times \hat{j}) = \hat{i} \times \hat{k} = -\hat{j}$

(b) $(\hat{i} \times \hat{j}) \times \hat{k} = \hat{k} \times \hat{k} = \vec{0}$

12.41. Solve: (a) $\vec{C} \times \vec{D} = 0$ implies that $\vec{D}$ must also be in the same or opposite direction as the $\vec{C}$ vector or zero, because $\hat{i} \times \hat{i} = 0$. Thus $\vec{D} = n\hat{i}$, where n could be any real number.

(b) $\vec{C} \times \vec{E} = 6\hat{k}$ implies that $\vec{E}$ must be along the $\hat{j}$ vector, because $\hat{i} \times \hat{j} = \hat{k}$. Thus $\vec{E} = 2\hat{j}$.

(c) $\vec{C} \times \vec{F} = -3\hat{j}$ implies that $\vec{F}$ must be along the $\hat{k}$ vector, because $\hat{i} \times \hat{k} = -\hat{j}$. Thus $\vec{F} = 1\hat{k}$.

12.43. Solve:
$$\vec{L} = \vec{r} \times m\vec{v} = (3.0\hat{i} + 2.0\hat{j}) \text{ m} \times (0.1 \text{ kg})(4.0\hat{j}) \text{ m/s}$$
$$= 1.20(\hat{i} \times \hat{j}) \text{ kg m}^2/\text{s} + 0.8(\hat{j} \times \hat{j}) \text{ kg m}^2/\text{s} = 1.20\hat{k} \text{ kg m}^2/\text{s} + 0 \text{ kg m}^2/\text{s}$$
$$= 1.20\hat{k} \text{ kg m}^2/\text{s or } (1.20 \text{ kg m}^2/\text{s, out of page})$$

Section 12.11 Angular Momentum

12.47. Model: The bowling ball is a solid sphere.
Solve: From Table 12.2, the moment of inertia about a diameter of a solid sphere is
$$I = \frac{2}{5}MR^2 = \frac{2}{5}(5.0 \text{ kg})(0.11 \text{ m})^2 = 0.0242 \text{ kg m}^2$$
Require
$$L = 0.23 \text{ kg m}^2/\text{s} = I\omega = (0.0243 \text{ kg m}^2)\omega$$
$$\Rightarrow \omega = (9.5 \text{ rad/s})$$
In rpm, this is $(9.5 \text{ rad/s})\left(\dfrac{\text{rev}}{2\pi \text{ rad}}\right)\left(\dfrac{60 \text{ s}}{\text{min}}\right) = 91 \text{ rpm}$.

12.51. Model: The wheel is a rigid rolling body.
Visualize:

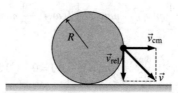

Solve: The front of the disk is moving forward at velocity v_{cm}. Also, because of rotation the point is moving downward at velocity $v_{\text{rel}} = R\omega = v_{\text{cm}}$. So, this point has a speed
$$v = \sqrt{v_{\text{cm}}^2 + v_{\text{cm}}^2} = \sqrt{2}v_{\text{cm}} = \sqrt{2}(20 \text{ m/s}) = 28 \text{ m/s}$$
Assess: The speed v is independent of the radius of the wheel.

12.53. Model: The disk is a rigid rotating body. The axis is perpendicular to the plane of the disk.
Visualize:

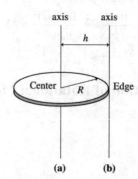

Solve: (a) From Table 12.2, the moment of inertia of a disk about its center is

$$I = \frac{1}{2}MR^2 = \frac{1}{2}(2.0\ \text{kg})(0.10\ \text{m})^2 = 0.010\ \text{kg m}^2$$

(b) To find the moment of inertia of the disk through the edge, we can make use of the parallel axis theorem:

$$I = I_{\text{center}} + Mh^2 = (0.010\ \text{kg m}^2) + (2.0\ \text{kg})(0.10\ \text{m})^2 = 0.030\ \text{kg m}^2$$

Assess: The larger moment of inertia about the edge means there is more inertia to rotational motion about the edge than about the center.

12.57. Model: The plate has uniform density.
Visualize:

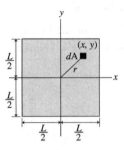

Solve: The moment of inertia is

$$I = \int r^2\,dm.$$

Let the mass of the plate be M. Its area is L^2. A region of area dA located at (x, y) has mass $dm = \frac{M}{A}dA = \frac{M}{L^2}dx\,dy$.

The distance from the axis of rotation to the point (x, y) is $r = \sqrt{x^2 + y^2}$. With $-\frac{L}{2} \le x \le \frac{L}{2}$ and $-\frac{L}{2} \le y \le \frac{L}{2}$,

$$I = \int_{-\frac{L}{2}}^{\frac{L}{2}} \int_{-\frac{L}{2}}^{\frac{L}{2}} (x^2 + y^2)\left(\frac{M}{L^2}\right)dx\,dy = \frac{M}{L^2}\int_{-\frac{L}{2}}^{\frac{L}{2}} \left(\frac{x^3}{3} + y^2 x\right)\Bigg|_{-\frac{L}{2}}^{\frac{L}{2}} dy$$

$$= \frac{M}{L^2}\int_{-\frac{L}{2}}^{\frac{L}{2}} \left(\frac{L^3}{24} + \frac{y^2 L}{2} - \left(\frac{-L^3}{24} - \frac{y^2 L}{2}\right)\right)dy = \frac{M}{L^2}\int_{-\frac{L}{2}}^{\frac{L}{2}} \left(\frac{L^3}{12} + Ly^2\right)dy = \frac{M}{L^2}\left(\frac{L^4}{12} + L\frac{y^3}{3}\right)\Bigg|_{-\frac{L}{2}}^{\frac{L}{2}}$$

$$= \frac{M}{L^2}\left(\frac{L^4}{12} + L\left(\frac{L^3}{24} + \frac{L^3}{24}\right)\right) = \frac{1}{6}ML^2$$

12.59. Model: Assume the woman is in equilibrium, so $\sum F = 0$ and $\sum \tau = 0$.
Visualize: Choose the axis to be the left end of the board.

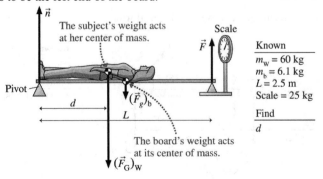

Solve: Use $\Sigma_\tau = 0$.

$$\Sigma_\tau = L(\text{scalereading})g - d(m_w)g - \frac{L}{2}(m_b)g = 0\,\text{N}\cdot\text{m}$$

$$d = \frac{L(\text{scalereading})g - \frac{L}{2}(m_b)g}{m_w g} = \frac{L[(\text{scalereading}) - \frac{1}{2}m_b]}{m_w} =$$

$$\frac{(2.5\,\text{m})[(25\,\text{kg}) - \frac{1}{2}(6.1\,\text{kg})]}{60\,\text{kg}} = 0.91\,\text{m}$$

Assess: This is a little more than halfway up the body of a woman of average height.

12.63. Model: Model the beam as a rigid body. For the beam not to fall over, it must be both in translational equilibrium ($\vec{F}_{\text{net}} = \vec{0}\,\text{N}$) and rotational equilibrium ($\tau_{\text{net}} = 0\,\text{Nm}$).
Visualize:

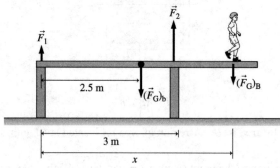

The boy walks along the beam a distance x, measured from the left end of the beam. There are four forces acting on the beam. F_1 and F_2 are from the two supports, $(\vec{F}_G)_b$ is the gravitational force on the beam, and $(\vec{F}_G)_B$ is the gravitational force on the boy.
Solve: We pick our pivot point on the left end through the first support. The equation for rotational equilibrium is
$$-(F_G)_b(2.5\,\text{m}) + F_2(3.0\,\text{m}) - (F_G)_B x = 0\,\text{N m}$$
$$-(40\,\text{kg})(9.80\,\text{m/s}^2)(2.5\,\text{m}) + F_2(3.0\,\text{m}) - (20\,\text{kg})(9.80\,\text{m/s}^2)x = 0\,\text{N m}$$
The equation for translation equilibrium is
$$\Sigma F_y = 0\,\text{N} = F_1 + F_2 - (F_G)_b - (F_G)_B$$
$$\Rightarrow F_1 + F_2 = (F_G)_b + (F_G)_B = (40\,\text{kg} + 20\,\text{kg})(9.8\,\text{m/s}^2) = 588\,\text{N}$$
Just when the boy is at the point where the beam tips, $F_1 = 0\,\text{N}$. Thus $F_2 = 588\,\text{N}$. With this value of F_2, we can simplify the torque equation to:
$$-(40\,\text{kg})(9.80\,\text{m/s}^2)(2.5\,\text{m}) + (588\,\text{N})(3.0\,\text{m}) - (20\,\text{kg})(9.80\,\text{m/s}^2)x = 0\,\text{N m}$$
$$\Rightarrow x = 4.0\,\text{m}$$
Thus, the distance from the right end is $5.0\,\text{m} - 4.0\,\text{m} = 1.0\,\text{m}$.

12.65. Model: The pole is a uniform rod. The sign is also uniform.
Visualize:

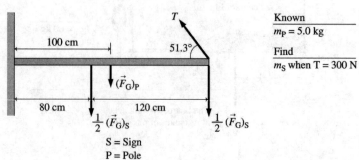

Solve: The geometry of the rod and cable give the angle that the cable makes with the rod.

$$\theta = \tan^{-1}\left(\frac{250}{200}\right) = 51.3°$$

The rod is in rotational equilibrium about its left-hand end.

$$\tau_{net} = 0 = -(100 \text{ cm})(F_G)_P - (80 \text{ cm})\left(\frac{1}{2}\right)(F_G)_S - (200 \text{ cm})\left(\frac{1}{2}\right)(F_G)_S + (200 \text{ cm})T\sin 51.3°$$

$$= -(100 \text{ cm})(5.0 \text{ kg})(9.8 \text{ m/s}^2) - m_S(9.8 \text{ m/s}^2)(140 \text{ cm}) + (156 \text{ cm})T$$

With $T = 300 \ N$, $m_S = 30.6 \text{ kg} \approx 31 \text{ kg}$.

Assess: A mass of 30.6 kg is reasonable for a sign.

12.67. Model: The bar is a solid body rotating through its center.
Visualize:

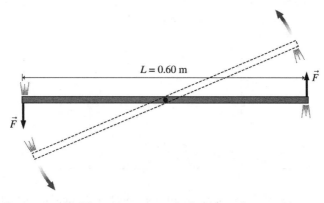

$L = 0.60$ m

Solve: **(a)** The two forces form a couple. The net torque on the bar about its center is

$$\tau_{net} = LF = I\alpha \Rightarrow F = \frac{I\alpha}{L}$$

where F is the force produced by one of the air jets. We can find I and α as follows:

$$I = \frac{1}{12}ML^2 = \frac{1}{12}(0.50 \text{ kg})(0.60 \text{ m})^2 = 0.015 \text{ kg m}^2$$

$$\omega_1 = \omega_0 + \alpha(t_1 - t_0) \Rightarrow 150 \text{ rpm} = 5.0\pi \text{ rad/s} = 0 \text{ rad} + \alpha(10 \text{ s} - 0 \text{ s}) \Rightarrow \alpha = 0.50\pi \text{ rad/s}^2$$

$$\Rightarrow F = \frac{(0.015 \text{ kg m}^2)(0.5\pi \text{ rad/s}^2)}{(0.60 \text{ m})} = 0.0393 \text{ N}$$

The force $F - 39$ mN.

(b) The torque of a couple is the same about any point. It is still $\tau_{net} = LF$. However, the moment of inertia has changed.

$$\tau_{net} = LF = I\alpha \Rightarrow \alpha = \frac{LF}{I} \text{ where } I = \frac{1}{3}ML^2 = \frac{1}{3}(0.500 \text{ kg})(0.6 \text{ m})^2 = 0.060 \text{ kg m}^2$$

$$\Rightarrow \alpha = \frac{(0.0393 \text{ N})\times(0.60 \text{ m})}{0.060 \text{ kg m}^2} = 0.393 \text{ rad/s}^2$$

Finally,

$$\omega_1 = \omega_0 + \alpha(t_1 - t_0) = 0 \text{ rad/s} + (0.393 \text{ rad/s}^2)(10 \text{ s} - 0 \text{ s})$$

$$= 3.93 \text{ rad/s} = \frac{(3.93)(60)}{2\pi} \text{ rpm} = 37.5 \text{ rpm}$$

The angular speed is 38 rpm.
Assess: Note that $\omega \propto \alpha$ and $\alpha \propto 1/I$. Thus, $\omega \propto 1/I$. I about the center of the rod is 4 times smaller than I about one end of the rod. Consequently, ω is 4 times larger.

12.71. Model: The disk is a rigid spinning body.
Visualize: Please refer to Figure P12.71. The initial angular velocity is 300 rpm or $(300)(2\pi)/60 = 10\pi$ rad/s. After 3.0 s the disk stops.
Solve: Using the kinematic equation for angular velocity,

$$\omega_1 = \omega_0 + \alpha(t_1 - t_0) \Rightarrow \alpha = \frac{\omega_1 - \omega_0}{t_1 - t_0} = \frac{(0 \text{ rad/s} - 10\pi \text{ rad/s})}{(3.0 \text{ s} - 0 \text{ s})} = \frac{-10\pi}{3} \text{ rad/s}^2$$

Thus, the torque due to the force of friction that brings the disk to rest is

$$\tau = I\alpha = -fR \Rightarrow f = -\frac{I\alpha}{R} = -\frac{(\frac{1}{2}mR^2)\alpha}{R} = -\frac{1}{2}(mR)\alpha = -\frac{1}{2}(2.0 \text{ kg})(0.15 \text{ m})\left(-10\frac{\pi}{3} \text{ rad/s}^2\right) = 1.57 \text{ N} \approx 1.6 \text{ N}$$

The minus sign with $\tau = -fR$ indicates that the torque due to friction acts clockwise.

12.73. Model: Assume that the hollow sphere is a rigid rolling body and that the sphere rolls up the incline without slipping. We also assume that the coefficient of rolling friction is zero.
Visualize:

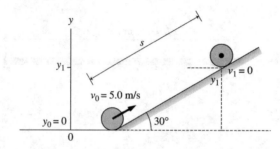

The initial kinetic energy, which is a combination of rotational and translational energy, is transformed in gravitational potential energy. We chose the bottom of the incline as the zero of the gravitational potential energy.
Solve: The conservation of energy equation $K_f + U_{gf} = K_i + U_{gi}$ is

$$\frac{1}{2}M(v_1)^2_{cm} + \frac{1}{2}I_{cm}(\omega_1)^2 + Mgy_1 = \frac{1}{2}M(v_0)^2_{cm} + \frac{1}{2}I_{cm}(\omega_0)^2 + Mgy_0$$

$$0 \text{ J} + 0 \text{ J} + Mgy_1 = \frac{1}{2}M(v_0)^2_{cm} + \frac{1}{2}\left(\frac{2}{3}MR^2\right)(\omega_0)^2_{cm} + 0 \text{ J} \Rightarrow Mgy_1 = \frac{1}{2}M(v_0)^2_{cm} + \frac{1}{3}MR^2\frac{(v_0)^2_{cm}}{R^2}$$

$$\Rightarrow gy_1 = \frac{5}{6}(v_0)^2_{cm} \Rightarrow y_1 = \frac{\frac{5}{6}(v_0)^2_{cm}}{g} = \frac{5}{6}\frac{(5.0 \text{ m/s})^2}{9.8 \text{ m/s}^2} = 2.126 \text{ m}$$

The distance traveled along the incline is

$$s = \frac{y_1}{\sin 30°} = \frac{2.126 \text{ m}}{0.5} = 4.3 \text{ m}$$

Assess: This is a reasonable stopping distance for an object rolling up an incline when its speed at the bottom of the incline is approximately 10 mph.

12.75. Model: The hoop is a rigid body rotating about an axle at the edge of the hoop. The gravitational torque on the hoop causes it to rotate, transforming the gravitational potential energy of the hoop's center of mass into rotational kinetic energy.

Visualize:

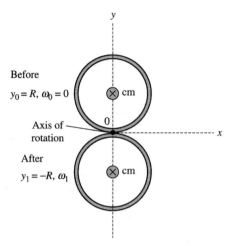

Before
$y_0 = R, \omega_0 = 0$

Axis of rotation

After
$y_1 = -R, \omega_1$

We placed the origin of the coordinate system at the hoop's edge on the axle. In the initial position, the center of mass is a distance R above the origin, but it is a distance R below the origin in the final position.

Solve: (a) Applying the parallel-axis theorem, $I_{edge} = I_{cm} + mR^2 = mR^2 + mR^2 = 2mR^2$. Using this expression in the energy conservation equation $K_f + U_{gf} = K_i + U_{gi}$ yields:

$$\frac{1}{2}I_{edge}\omega_1^2 + mgy_1 = \frac{1}{2}I_{edge}\omega_0^2 + mgy_0 \qquad \frac{1}{2}(2mR^2)\omega_1^2 - mgR = 0\text{ J} + mgR \Rightarrow \omega_1 = \sqrt{\frac{2g}{R}}$$

(b) The speed of the lowest point on the hoop is

$$v = (\omega_1)(2R) = \sqrt{\frac{2g}{R}}(2R) = \sqrt{8gR}$$

Assess: Note that the speed of the lowest point on the loop involves a distance of $2R$ instead of R.

12.81. Model: Model the merry-go-round as a rigid disk rotating on frictionless bearings about an axle in the center and John as a particle. For the (merry-go-round + John) system, no external torques act as John jumps on the merry-go-round. Angular momentum is thus conserved.

Visualize: The initial angular momentum is the sum of the angular momentum of the merry-go-round and the angular momentum of John. The final angular momentum as John jumps on the merry-go-round is equal to $I_{final}\omega_{final}$.

Solve: John's initial angular momentum is that of a particle: $L_J = m_J v_J R \sin\beta = m_J v_J R$. The angle $\beta = 90°$ since John runs tangent to the disk. The conservation of angular momentum equation $L_f - L_i$ is

$$I_{final}\omega_{final} = I_{disk} + I_J = \left(\frac{1}{2}MR^2\right)\omega_i + m_J v_J R$$

$$= \left(\frac{1}{2}\right)(250\text{ kg})(1.5\text{ m})^2(20\text{ rpm})\frac{2\pi}{60}\left(\frac{\text{rad}}{\text{rpm}}\right) + (30\text{ kg})(5.0\text{ m/s})(1.5\text{ m}) = 814\text{ kg m}^2/\text{s}$$

$$\Rightarrow \omega_{final} = \frac{814\text{ kg m}^2/\text{s}}{I_{final}}$$

$$I_{final} = I_{disk} + I_J = \frac{1}{2}MR^2 + m_J R^2 = \frac{1}{2}(250\text{ kg})(1.5\text{ m})^2 + (30\text{ kg})(1.5\text{ m})^2 = 349\text{ kg m}^2$$

$$\omega_{final} = \frac{814\text{ kg m}^2/\text{s}}{349\text{ kg m}^2} = 2.33\text{ rad/s} = 22\text{ rpm}$$

NEWTON'S THEORY OF GRAVITY

Exercises and Problems

Section 13.3 Newton's Law of Gravity

13.1. Model: Model the sun (s) and the earth (e) as spherical masses. Due to the large difference between your size and mass and that of either the sun or the earth, a human body can be treated as a particle.

Solve: $F_{\text{s on you}} = \dfrac{GM_sM_y}{r_{\text{s-e}}^2}$ and $F_{\text{e on you}} = \dfrac{GM_eM_y}{r_e^2}$

Dividing these two equations gives

$$\frac{F_{\text{s on y}}}{F_{\text{e on y}}} = \left(\frac{M_s}{M_e}\right)\left(\frac{r_e}{r_{\text{s-e}}}\right)^2 = \left(\frac{1.99\times10^{30}\text{ kg}}{5.98\times10^{24}\text{ kg}}\right)\left(\frac{6.37\times10^6\text{ m}}{1.50\times10^{11}\text{ m}}\right)^2 = 6.00\times10^{-4}$$

13.5. Model: Model the woman (w) and the man (m) as spherical masses or particles.

Solve: $F_{\text{w on m}} = F_{\text{m on w}} = \dfrac{GM_wM_m}{r_{\text{m-w}}^2} = \dfrac{(6.67\times10^{-11}\text{ N}\cdot\text{m}^2/\text{kg}^2)(50\text{ kg})(70\text{ kg})}{(1.0\text{ m})^2} = 2.3\times10^{-7}\text{ N}$

Section 13.4 Little *g* and Big *G*

13.9. Model: Model the earth (e) as a spherical mass.
Visualize: The acceleration due to gravity at sea level is 9.83 m/s² (see Table 13.1) and $R_e = 6.37\times10^6$ m (see Table 13.2).

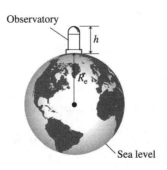

Observatory

h

R_e

Sea level

Solve: $g_{\text{observatory}} = \dfrac{GM_e}{(R_e + h)^2} = \dfrac{GM_e}{R_e^2 (1 + h/R_e)^2} = \dfrac{g_{\text{earth}}}{(1 + h/R_e)^2} = (9.83 - 0.0075)\ \text{m/s}^2$

Here $g_{\text{earth}} = GM_e/R_e^2$ is the acceleration due to gravity on a non-rotating earth, which is why we've used the value $9.83\ \text{m/s}^2$. Solving for h,

$$h = \left(\sqrt{\dfrac{9.83}{9.8225}} - 1 \right) R_e = 2.4\ \text{km}$$

13.11. Model: Model Planet Z as a spherical mass.

Solve: (a) $g_{Z\ \text{surface}} = \dfrac{GM_Z}{R_Z^2} \Rightarrow 8.0\ \text{m/s}^2 = \dfrac{(6.67 \times 10^{-11}\ \text{N} \cdot \text{m}^2/\text{kg}^2)M_Z}{(5.0 \times 10^6\ \text{m})^2} \Rightarrow M_Z = 3.0 \times 10^{24}\ \text{kg}$

(b) Let h be the height above the north pole. Thus,

$$g_{\text{above N pole}} = \dfrac{GM_Z}{(R_Z + h)^2} = \dfrac{GM_Z}{R_Z^2 (1 + h/R_Z)^2} = \dfrac{g_{Z\ \text{surface}}}{(1 + h/R_Z)^2} = \dfrac{8.0\ \text{m/s}^2}{\left(1 + \dfrac{10.0 \times 10^6\ \text{m}}{5.0 \times 10^6\ \text{m}} \right)^2} = 0.89\ \text{m/s}^2$$

Section 13.5 Gravitational Potential Energy

13.13. Model: Model Jupiter as a spherical mass and the object as a point particle. The object and Jupiter form an isolated system, so mechanical energy is conserved. The minimum launch speed for escape, which is called the escape speed, allows an object to escape to an infinite distance from Jupiter (or, in general, from its partner object).
Visualize:

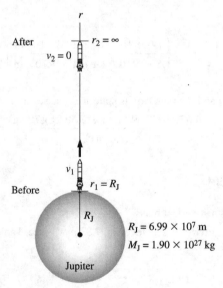

Solve: The energy conservation equation $K_2 + U_2 = K_1 + U_1$ is

$$\tfrac{1}{2} m_o v_2^2 - \dfrac{GM_J m_o}{r_2} = \tfrac{1}{2} m_o v_1^2 - \dfrac{GM_J m_o}{R_J}$$

where R_J and M_J are the radius and mass of Jupiter. Using the asymptotic condition $v_2 = 0\ \text{m/s}$ as $r_2 \to \infty$,

$$0\ \text{J} = \tfrac{1}{2} m_o v_1^2 - \dfrac{GM_J m_o}{R_J} \Rightarrow v_1 = \sqrt{\dfrac{2GM_J}{R_J}} = \sqrt{\dfrac{2(6.67 \times 10^{-11}\ \text{N} \cdot \text{m}^2/\text{kg}^2)(1.90 \times 10^{27}\ \text{kg})}{6.99 \times 10^7\ \text{m}}} = 6.02 \times 10^4\ \text{m/s}$$

Thus, the escape velocity from Jupiter is 60.2 km/s.

Section 13.6 Satellite Orbits and Energies

13.17. Model: Model the sun (s) as a spherical mass and the asteroid (a) as a point particle.
Visualize: The asteroid, having mass m_a and velocity v_a, orbits the sun in a circle of radius r_a. The asteroid's time period is $T_a = 5.0$ earth years $= 1.58 \times 10^8$ s.
Solve: The gravitational force between the sun (mass $= M_s$) and the asteroid provides the centripetal acceleration required for circular motion.

$$\frac{GM_s m_a}{r_a^2} = \frac{m_a v_a^2}{r_a} \implies \frac{GM_s}{r_a} = \left(\frac{2\pi r_a}{T_a}\right)^2 \implies r_a = \left(\frac{GM_s T_a^2}{4\pi^2}\right)^{1/3}$$

Substituting $G = 6.67 \times 10^{-11}$ N·m^2/kg^2, $M_s = 1.99 \times 10^{30}$ kg, and the time period of the asteroid, we obtain $r_a = 4.4 \times 10^{11}$ m. The velocity of the asteroid in its orbit will therefore be

$$v_a = \frac{2\pi r_a}{T_a} = \frac{2\pi(4.4 \times 10^{11} \text{ m})}{1.58 \times 10^8 \text{ s}} = 1.7 \times 10^4 \text{ m/s}$$

Solve: We give the answer to two significant figures because the asteroid period is given to two significant figures.

13.19. From Kepler's third law, the orbital period squared is proportional to the orbital radius cubed: $T^2 \propto r^3$. Thus, at $r_y = 4r_x$, $T_y^2 \propto (4r_x)^3 = 64r_x^3 \propto (8T_x)^2$. Therefore $T_y = 8T_x$. A year on Planet Y is $8 \times 200 = 1600$ earth days long.

13.23. Model: Model the earth (e) as a spherical mass and the satellite (s) as a point particle.
Visualize: The satellite has a mass is m_s and orbits the earth with a velocity v_s. The radius of the circular orbit is denoted by r_s and the mass of the earth by M_e.
Solve: The satellite experiences a gravitational force that provides the centripetal acceleration required for circular motion:

$$\frac{GM_e m_s}{r_s^2} = \frac{m_s v_s^2}{r_s} \implies r_s = \frac{GM_e}{v_s^2} = \frac{(6.67 \times 10^{-11} \text{ N·m}^2/\text{kg}^2)(5.98 \times 10^{24} \text{ kg})}{(5500 \text{ m/s})^2} = 1.32 \times 10^7 \text{ m}$$

$$T_s = \frac{2\pi R_s}{v_s} = \frac{2\pi(1.32 \times 10^7 \text{ m})}{5500 \text{ m/s}} = 1.51 \times 10^4 \text{ s} = 4.2 \text{ h}$$

13.27. Visualize:

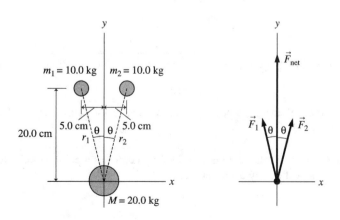

Solve: The angle $\theta = \tan^{-1}\left(\dfrac{5}{20}\right) = 14.04°$. The distance $r_1 = r_2 = \sqrt{(0.050 \text{ m})^2 + (0.200 \text{ m})^2} = 0.206$ m. The forces on the 20.0 kg mass are

$$\vec{F}_1 = G\frac{Mm_1}{r_1^2}(-\sin\theta\hat{i} + \cos\theta\hat{j})$$

$$\vec{F}_2 = G\frac{Mm_2}{r_2^2}(\sin\theta\hat{i} + \cos\theta\hat{j})$$

Note $m_1 = m_2$ and $r_1 = r_2$. Thus, the net force on the 20.0 kg mass is

$$\vec{F}_{net} = \vec{F}_1 + \vec{F}_2 = 2G\frac{Mm_1}{r_1^2}\cos\theta\hat{j}$$

$$= \frac{2(6.67\times10^{-11} \text{ N}\cdot\text{m}^2/\text{kg}^2)(20.0 \text{ kg})(5.0 \text{ kg})\cos(14.04°)}{(0.206 \text{ m})^2}\hat{j}$$

$$= 3.0\times10^{-7}\,\hat{j}\text{ N}$$

13.29. Visualize:

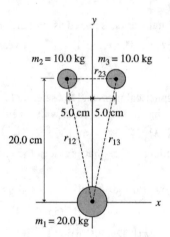

Solve: The total gravitational potential energy is the sum of the potential energies due to the interactions of the pairs of masses.

$$U = U_{12} + U_{13} + U_{23}$$

$$= -G\frac{m_1m_2}{r_{12}} - G\frac{m_1m_3}{r_{13}} - G\frac{m_2m_3}{r_{23}}$$

With $m_1 = 20.0$ kg, $m_2 = m_3 = 10.0$ kg, $r_{12} = r_{13} = \sqrt{(0.050 \text{ m})^2 + (0.200 \text{ m})^2} = 0.206$ m, and $r_{23} = 0.100$ m,

$$U = -1.96\times10^{-7} \text{ J}$$

Assess: The gravitational potential energy is negative because the masses attract each other. It is a scalar, so there are no vector calculations required.

13.33. Model: Model the earth as a spherical mass and the object (o) as a point particle. Ignore air resistance. This is an isolated system, so mechanical energy is conserved.

Visualize:

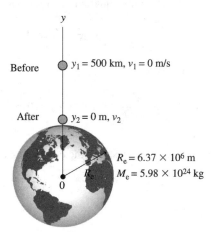

Solve: **(a)** The conservation of energy equation $K_2 + U_{g2} = K_1 + U_{g1}$ is

$$\frac{1}{2}m_o v_2^2 - \frac{GM_e m_o}{R_e} = \frac{1}{2}m_o v_1^2 - \frac{GM_e m_o}{R_e + y_1}$$

$$v_2 = \sqrt{2GM_e\left(\frac{1}{R_e} - \frac{1}{R_e + y_1}\right)}$$

$$= \sqrt{2(6.67 \times 10^{-11} \text{ N} \cdot \text{m}^2/\text{kg}^2)(5.98 \times 10^{24} \text{ kg})\left(\frac{1}{6.37 \times 10^6 \text{ m}} - \frac{1}{6.87 \times 10^6 \text{ m}}\right)} = 3.02 \text{ km/s}$$

(b) In the flat-earth approximation, $U_g = mgy$. The energy conservation equation thus becomes

$$\frac{1}{2}m_o v_2^2 + m_o g y_2 = \frac{1}{2}m_o v_1^2 + m_o g y_1$$

$$v_2 = \sqrt{v_1^2 + 2g(y_1 - y_2)} = \sqrt{2(9.81 \text{ m/s}^2)(5.00 \times 10^5 \text{ m} - 0 \text{ m})} = 3.13 \text{ km/s}$$

(c) The percent error in the flat-earth calculation is

$$\frac{3130 \text{ m/s} - 3020 \text{ m/s}}{3020 \text{ m/s}} \approx 3.6\%$$

13.35. Model: Model the earth and the projectile as spherical masses. Ignore air resistance. This is an isolated system, so mechanical energy is conserved.
Visualize:

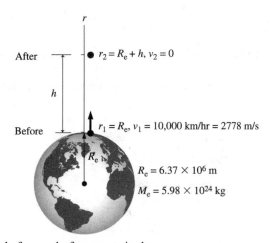

A pictorial representation of the before-and-after events is shown.

Solve: After using $v_2 = 0$ m/s, the energy conservation equation $K_2 + U_2 = K_1 + U_1$ is

$$0 \text{ J} - \frac{GM_e m_p}{R_e + h} = \frac{1}{2} m_p v_1^2 - \frac{GM_e m_p}{R_e}$$

The projectile mass cancels. Solving for h, we find

$$h = \left[\frac{1}{R_e} - \frac{v_1^2}{2GM_e}\right]^{-1} - R_e = 4.2 \times 10^5 \text{ m}$$

13.37. Model: Model the two stars as spherical masses, and the comet as a point mass. This is an isolated system, so mechanical energy is conserved.
Visualize:

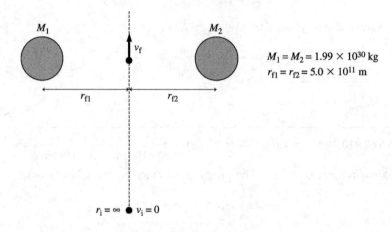

In the initial state, the comet is far away from the two stars and thus it has neither kinetic energy nor potential energy. In the final state, as the comet passes through the midpoint connecting the two stars, it possesses both kinetic energy and potential energy.
Solve: The conservation of energy equation $K_f + U_f = K_i + U_i$ gives

$$\frac{1}{2} m v_f^2 - \frac{GMm}{r_{f1}} - \frac{GMm}{r_{f2}} = 0 \text{ J} + 0 \text{ J}$$

$$v_f = \sqrt{\frac{4GM}{r_f}} = \sqrt{\frac{4(6.67 \times 10^{-11} \text{ N} \cdot \text{m}^2/\text{kg}^2)(1.99 \times 10^{30} \text{ kg})}{0.50 \times 10^{12} \text{ m}}} = 32{,}600 \text{ m/s} = 33 \text{ km/s}$$

Assess: Note that the final velocity of 33 km/s does not depend on the mass of the comet.

13.41. Model: Gravity is a conservative force, so we can use conservation of energy.
Visualize:

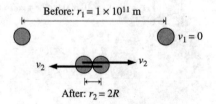

The planets will be pulled together by gravity and each will have speed v_2 as they crash and the separation between their centers will be $2R$.

Solve: The planets begin with only gravitational potential energy. When they crash, they have both potential and kinetic energy. Thus,

$$K_2 + U_2 = \frac{1}{2}Mv_2^2 + \frac{1}{2}Mv_2^2 - \frac{GMM}{r_2} = K_1 + U_1 = 0 \text{ J} - \frac{GMM}{r_1}$$

$$v_2 = \sqrt{GM\left(\frac{1}{r_2} - \frac{1}{r_1}\right)}$$

Because the planet is "Jupiter-size," we'll use $M = M_{Jupiter} = 1.9 \times 10^{27}$ kg and $r_2 = 2R_{Jupiter} = 1.4 \times 10^8$ m. Inserting these values into the expression above gives the crash speed of each planet as $v_2 = 3.0 \times 10^4$ m/s.

Assess: Note that the force is not constant, because it varies with distance, so the motion is *not* constant acceleration motion. The formulas from constant-acceleration kinematics do not work for problems such as this.

13.45. Model: Model the moon (m) as a spherical mass and the lander (l) as a particle. This is an isolated system, so mechanical energy is conserved.

Visualize: The initial position of the lunar lander (mass $= m_l$) is at a distance $r_1 = R_m + 50$ km from the center of the moon. The final position of the lunar lander is the orbit whose distance from the center of the moon is $r_2 = R_m + 300$ km.

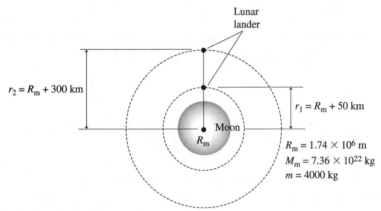

Solve: The external work done by the thrusters is

$$W_{ext} = \Delta E_{mech} = \tfrac{1}{2}\Delta U_g$$

where we used $E_{mech} = \tfrac{1}{2}U_g$ for a circular orbit. The change in potential energy is from the initial orbit at $r_i = R_m + 50$ km to the final orbit $r_f = R_m + 300$ km. Thus

$$W_{ext} = \frac{1}{2}\left(\frac{-GM_m m}{r_f} - \frac{-GM_m m}{r_i}\right) = \frac{GM_m m}{2}\left(\frac{1}{r_i} - \frac{1}{r_f}\right)$$

$$= \frac{(6.67\times10^{-11} \text{ N}\cdot\text{m}^2/\text{kg}^2)(7.36\times10^{22} \text{ kg})(4000 \text{ kg})}{2}\left(\frac{1}{1.79\times10^6 \text{ m}} - \frac{1}{2.04\times10^6 \text{ m}}\right)$$

$$= 6.7\times10^8 \text{ J}$$

13.47. Model: Assume a spherical asteroid and a point mass model for the satellite. This is an isolated system, so mechanical energy is conserved.

Visualize:

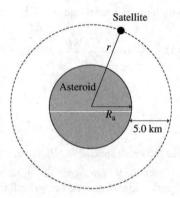

The orbital radius of the satellite is

$$r = R_a + h = 8{,}800 \text{ m} + 5{,}000 \text{ m} = 13{,}800 \text{ m}$$

Solve: (a) The speed of a satellite in a circular orbit is

$$v = \sqrt{\frac{GM}{r}} = \left[\frac{(6.67 \times 10^{-11} \text{ N} \cdot \text{m}^2/\text{kg}^2)(1.0 \times 10^{16} \text{ kg})}{13{,}800 \text{ m}} \right]^{1/2} = 7.0 \text{ m/s}$$

(b) The minimum launch speed for escape (v_i) will cause the satellite to stop asymptotically ($v_f = 0$ m/s) as $r_f \to \infty$. Using the energy conservation equation $K_2 + U_2 = K_1 + U_1$, we get

$$\frac{1}{2}m_s v_f^2 - \frac{GM_a m_s}{r_f} = \frac{1}{2}m_s v_i^2 - \frac{GM_a m_s}{R_a} \quad \Rightarrow \quad 0 \text{ J} - 0 \text{ J} = \frac{1}{2}v_{escape}^2 - \frac{GM_a}{R_a}$$

$$v_{escape} = \sqrt{\frac{2GM_a}{R_a}} = \sqrt{\frac{2(6.67 \times 10^{-11} \text{ N} \cdot \text{m}^2/\text{kg}^2)(1.0 \times 10^{16} \text{ kg})}{8800 \text{ m}}} = 12 \text{ m/s}$$

13.49. Model: Model the earth as a spherical mass and the satellite as a point mass.

Visualize: The satellite is directly over a point on the equator once every two days. Thus, $T = 2T_e = 2 \times 24 \times 3600 \text{ s} = 1.728 \times 10^5 \text{ s}$.

Solve: From Kepler's laws applied to a circular orbit:

$$T^2 = \left(\frac{4\pi^2}{GM_e} \right) r^3$$

$$r^3 = \frac{GM_e T^2}{4\pi^2} = \frac{(6.67 \times 10^{-11} \text{ N} \cdot \text{m}^2/\text{kg}^2)(5.98 \times 10^{24} \text{ kg})(1.728 \times 10^5)^2}{4\pi^2}$$

$$r = 6.71 \times 10^7 \text{ m}$$

Assess: The radius of the orbit is larger than the geosynchronous orbit.

13.51. Solve: (a) Taking the logarithm of both sides of $v^p = Cu^q$ gives

$$[\log(v^p) = p \log v] = [\log(Cu^q) = \log C + q \log u] \quad \Rightarrow \quad \log v = \frac{q}{p}\log u + \frac{\log C}{p}$$

But $x = \log u$ and $y = \log v$, so x and y are related by

$$y = \left(\frac{q}{p} \right) x + \frac{\log C}{p}$$

(b) The previous result shows there is a linear relationship between x and y, so there is a linear relationship between $\log u$ and $\log v$. The graph of a linear relationship is a straight line, so the graph of $\log v$-versus-$\log u$ will be a straight line.

(c) The slope of the straight line represented by the equation $y = (q/p)x + \log C/p$ is q/p. Thus, the slope of the $\log v$-versus-$\log u$ graph will be q/p.

(d) The predicted y-intercept of the graph is $\log C/p$, and the experimentally determined value is 9.264. Equating these, we can solve for M. Because the planets all orbit the sun, the mass we are finding is $M = M_{\text{sun}}$.

$$\frac{1}{2}\log C = \frac{1}{2}\log\left(\frac{4\pi^2}{GM_{\text{sun}}}\right) = -9.264 \quad \Rightarrow \quad \frac{4\pi^2}{GM_{\text{sun}}} = 10^{-18.528} = \frac{1}{10^{18.528}}$$

$$M_{\text{sun}} = \left(\frac{4\pi^2}{G}\right)(10^{18.528}) = 1.996\times10^{30} \text{ kg}$$

The tabulated value, to three significant figures, is $M_{\text{sum}} = 1.99\times10^{30}$ kg. We have used the orbits of the planets to "weigh the sun!"

13.57. Model: For the sun + comet system, the mechanical energy is conserved.

Visualize:

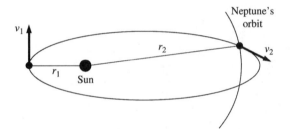

Solve: The conservation of energy equation $K_f + U_f = K_i + U_i$ gives

$$\frac{1}{2}M_c v_2^2 - \frac{GM_s M_c}{r_2} = \frac{1}{2}M_c v_1^2 - \frac{GM_s M_c}{r_1}$$

Using $G = 6.67\times10^{-11}$ Nm^2/kg^2, $M_s = 1.99\times10^{30}$ kg, $r_1 = 8.79\times10^{10}$ m, $r_2 = 4.50\times10^{12}$ m, and $v_1 = 54.6$ km/s, we get $v_2 = 4.49$ km/s.

OSCILLATIONS

Exercises and Problems

Section 14.1 Simple Harmonic Motion

14.3. Model: The air-track glider attached to a spring is in simple harmonic motion.
Visualize: The position of the glider can be represented as $x(t) = A\cos\omega t$.
Solve: The glider is pulled to the right and released from rest at $t = 0$ s. It then oscillates with a period $T = 2.0$ s and a maximum speed $v_{max} = 40$ cm/s $= 0.40$ m/s.

(a) $v_{max} = \omega A$ and $\omega = \dfrac{2\pi}{T} = \dfrac{2\pi}{2.0\text{ s}} = \pi$ rad/s $\Rightarrow A = \dfrac{v_{max}}{\omega} = \dfrac{0.40\text{ m/s}}{\pi\text{ rad/s}} = 0.127$ m $= 12.7$ cm $= 13$ cm

(b) The glider's position at $t = 0.25$ s is

$$x_{0.25\text{ s}} = (0.127\text{ m})\cos[(\pi\text{ rad/s})(0.25\text{ s})] = 0.090\text{ m} = 9.0\text{ cm}$$

Section 14.2 Simple Harmonic Motion and Circular Motion

14.5. Model: The oscillation is the result of simple harmonic motion.
Solve: **(a)** The amplitude $A = 10$ cm.
(b) The time to complete one cycle is the period, hence $T = 2.0$ s and

$$f = \frac{1}{T} = \frac{1}{2.0\text{ s}} = 0.50\text{ Hz}$$

(c) The position of an object undergoing simple harmonic motion is $x(t) = A\cos(\omega t + \phi_0)$. At $t = 0$ s, $x_0 = -5$ cm, thus

$$-5\text{ cm} = (10\text{ cm})\cos[\omega(0\text{ s}) + \phi_0]$$

$$\Rightarrow \cos\phi_0 = \frac{-5\text{ cm}}{10\text{ cm}} = -\frac{1}{2} \Rightarrow \phi_0 = \cos^{-1}\left(-\frac{1}{2}\right) = \pm\frac{2\pi}{3}\text{ rad or }\pm 120°$$

Since the oscillation is originally moving to the left, $\phi_0 = +120°$.

14.7. Visualize: A phase constant of $-\dfrac{\pi}{2}$ implies that the object that undergoes simple harmonic motion is in the lower half of the circular motion diagram. That is, the object is moving to the right.
Solve: The position of the object is given by the equation

$$x(t) = A\cos(\omega t + \phi_0) = A\cos(2\pi ft + \phi_0) = (8.0\text{ cm})\cos\left[\left(\frac{\pi}{2}\text{ rad/s}\right)t - \frac{\pi}{2}\text{ rad}\right]$$

The amplitude is $A = 8.0$ cm and the period is $T = 1/f = 4.0$ s. With $\phi_0 = -\pi/2$ rad, x starts at 0 cm and is moving to the right (getting more positive).

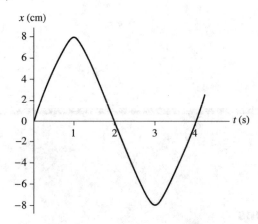

Assess: As we see from the graph, the object starts out moving to the right.

14.9. Solve: The position of the object is given by the equation

$$x(t) = A\cos(\omega t + \phi_0)$$

The amplitude is $A = 8.0$ cm. The angular frequency $\omega = 2\pi f = 2\pi(0.50 \text{ Hz}) = \pi$ rad/s. Since at $t = 0$ it has its most negative position, it must be about to move to the right, so $\phi_0 = -\pi$. Thus

$$x(t) = (8.0 \text{ cm})\cos[(\pi \text{ rad/s})t - \pi \text{ rad}]$$

Section 14.3 Energy in Simple Harmonic Motion

Section 14.4 The Dynamics of Simple Harmonic Motion

14.13. Model: The mass attached to the spring oscillates in simple harmonic motion.
Solve: (a) The period $T = 1/f = 1/2.0$ Hz $= 0.50$ s.

(b) The angular frequency $\omega = 2\pi f = 2\pi(2.0 \text{ Hz}) = 4\pi$ rad/s.

(c) Using energy conservation

$$\tfrac{1}{2}kA^2 = \tfrac{1}{2}kx_0^2 + \tfrac{1}{2}mv_{0x}^2$$

Using $x_0 = 5.0$ cm, $v_{0x} = -30$ cm/s and $k = m\omega^2 = (0.200 \text{ kg})(4\pi \text{ rad/s})^2$, we get $A = 5.54$ cm.

(d) To calculate the phase constant ϕ_0,

$$A\cos\phi_0 = x_0 = 5.0 \text{ cm}$$

$$\Rightarrow \phi_0 = \cos^{-1}\left(\frac{5.0 \text{ cm}}{5.54 \text{ cm}}\right) = 0.45 \text{ rad}$$

(e) The maximum speed is $v_{max} = \omega A = (4\pi \text{ rad/s})(5.54 \text{ cm}) = 70$ cm/s.

(f) The maximum acceleration is

$$a_{max} = \omega^2 A = \omega(\omega A) = (4\pi \text{ rad/s})(70 \text{ cm/s}) = 8.8 \text{ m/s}^2$$

(g) The total energy is $E = \tfrac{1}{2}mv_{max}^2 = \tfrac{1}{2}(0.200 \text{ kg})(0.70 \text{ m/s})^2 = 0.049$ J.

(h) The position at $t = 0.40$ s is

$$x_{0.4 \text{ s}} = (5.54 \text{ cm})\cos[(4\pi \text{ rad/s})(0.40 \text{ s}) + 0.45 \text{ rad}] = +3.8 \text{ cm}$$

Section 14.5 Vertical Oscillations

14.17. Model: The vertical oscillations constitute simple harmonic motion.
Visualize:

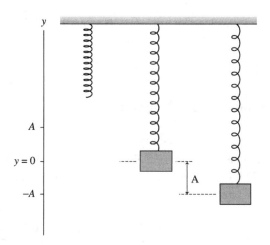

Solve: The period and angular frequency are

$$T = \frac{20 \text{ s}}{30 \text{ oscillations}} = 0.6667 \text{ s and } \omega = \frac{2\pi}{T} = \frac{2\pi}{0.6667 \text{ s}} = 9.425 \text{ rad/s}$$

(a) The mass can be found as follows:

$$\omega = \sqrt{\frac{k}{m}} \Rightarrow m = \frac{k}{\omega^2} = \frac{15 \text{ N/m}}{(9.425 \text{ rad/s})^2} = 0.169 \text{ kg} \approx 0.17 \text{ kg}$$

(b) The maximum speed $v_{max} = \omega A = (9.425 \text{ rad/s})(0.060 \text{ m}) = 0.57$ m/s.

Section 14.6 The Pendulum

14.19. Model: Assume a small angle of oscillation so there is simple harmonic motion.
Solve: The period of the pendulum is

$$T_0 = 2\pi\sqrt{\frac{L_0}{g}} = 4.0 \text{ s}$$

(a) The period is independent of the mass and depends only on the length. Thus $T = T_0 = 4.0$ s.
(b) For a new length $L = 2L_0$,

$$T = 2\pi\sqrt{\frac{2L_0}{g}} = \sqrt{2}T_0 = 5.7 \text{ s}$$

(c) For a new length $L = L_0/2$,

$$T = 2\pi\sqrt{\frac{L_0/2}{g}} = \frac{1}{\sqrt{2}}T_0 = 2.8 \text{ s}$$

(d) The period is independent of the amplitude as long as there is simple harmonic motion. Thus $T = 4.0$ s.

14.21. Model: Assume a small angle of oscillation so there is simple harmonic motion.
Solve: (a) On the earth the period is

$$T_{earth} = 2\pi\sqrt{\frac{L}{g}} = 2\pi\sqrt{\frac{1.0 \text{ m}}{9.80 \text{ m/s}^2}} = 2.0 \text{ s}$$

(b) On Venus the acceleration due to gravity is

$$g_{\text{Venus}} = \frac{GM_{\text{Venus}}}{R_{\text{Venus}}^2} = \frac{(6.67 \times 10^{-11} \text{ N} \cdot \text{m}^2/\text{kg}^2)(4.88 \times 10^{24} \text{ kg})}{(6.06 \times 10^6 \text{ m})^2} = 8.86 \text{ m/s}^2$$

$$\Rightarrow T_{\text{Venus}} = 2\pi \sqrt{\frac{L}{g_{\text{Venus}}}} = 2\pi \sqrt{\frac{1.0 \text{ m}}{8.86 \text{ m/s}^2}} = 2.1 \text{ s}$$

Section 14.7 Damped Oscillations

Section 14.8 Driven Oscillations and Resonance

14.25. Model: The spider is in simple harmonic motion.
Solve: Your tapping is a driving frequency. Largest amplitude at $f_{\text{ext}} = 1.0$ Hz means that this is the resonance frequency, so $f_0 = f_{\text{ext}} = 1.0$ Hz. That is, the spider's natural frequency of oscillation f_0 is 1.0 Hz and $\omega_0 = 2\pi f_0 = 2\pi$ rad/s. We have

$$\omega_0 = \sqrt{\frac{k}{m}} \Rightarrow k = m\omega_0^2 = (0.0020 \text{ kg})(2\pi \text{ rad/s})^2 = 0.079 \text{ N/m}$$

14.29. Model: Assume the eye is a simple driven oscillator.
Visualize: Given the mass and the resonant frequency, we can determine the effective spring constant using the relationship $\omega = 2\pi f = \sqrt{k/m}$.
Solve: Solving the above expression for the spring constant, obtain

$$k = (2\pi f)^2 m = [2\pi (29 \text{ Hz})]^2 (7.5 \times 10^{-3} \text{ kg}) = 249 \text{ N/m} \approx 250 \text{ N/m}$$

Assess: As spring constants go, this is a fairly large value, however the musculature holding the eyeball in the socket is strong and hence will have a large effective spring constant.

14.33. Solve: The object's position as a function of time is $x(t) = A\cos(\omega t + \phi_0)$. Letting $x = 0$ m at $t = 0$ s, gives

$$0 = A\cos\phi_0 \Rightarrow \phi_0 = \pm\tfrac{1}{2}\pi$$

Since the object is traveling to the right, it is in the lower half of the circular motion diagram, giving a phase constant between $-\pi$ and 0 radians. Thus, $\phi_0 = -\tfrac{1}{2}\pi$ and

$$x(t) = A\cos(\omega t - \tfrac{1}{2}\pi) \Rightarrow x(t) = A\sin\omega t = (0.10 \text{ m})\sin(\tfrac{1}{2}\pi t)$$

where we have used $A = 0.10$ m and

$$\omega = \frac{2\pi}{T} = \frac{2\pi \text{ rad}}{4.0 \text{ s}} = \frac{\pi}{2} \text{ rad/s}$$

Let us now find t where $x = 0.060$ m:

$$0.060 \text{ m} = (0.10 \text{ m})\sin\left(\frac{\pi}{2}t\right) \Rightarrow t = \frac{2}{\pi}\sin^{-1}\left(\frac{0.060 \text{ m}}{0.10 \text{ m}}\right) = 0.41 \text{ s}$$

Assess: The answer is reasonable because it is approximately $\tfrac{1}{8}$ of the period.

14.35. Model: The astronaut attached to the spring is in simple harmonic motion.
Solve: (a) From the graph, $T = 3.0$ s, so we have

$$T = 2\pi\sqrt{\frac{m}{k}} \Rightarrow m = \left(\frac{T}{2\pi}\right)^2 k = \left(\frac{3.0 \text{ s}}{2\pi}\right)^2 (240 \text{ N/m}) = 55 \text{ kg}$$

(b) Oscillations occur about an equilibrium position of 1.0 m. From the graph, $A = \tfrac{1}{2}(0.80 \text{ m}) = 0.40$ m, $\phi_0 = 0$ rad, and

$$\omega = \frac{2\pi}{T} = \frac{2\pi}{3.0 \text{ s}} = 2.1 \text{ rad/s}$$

The equation for the position of the astronaut is
$$x(t) = A\cos\omega t + 1.0 \text{ m} = (0.4 \text{ m})\cos[(2.1 \text{ rad/s})t] + 1.0 \text{ m}$$
$$\Rightarrow 1.2 \text{ m} = (0.4 \text{ m})\cos[(2.1 \text{ rad/s})t] + 1.0 \text{ m} \Rightarrow \cos[(2.1 \text{ rad/s})t] = 0.5 \Rightarrow t = 0.50 \text{ s}$$
The equation for the velocity of the astronaut is
$$v_x(t) = -A\omega\sin(\omega t)$$
$$\Rightarrow v_{0.5 \text{ s}} = -(0.4 \text{ m})(2.1 \text{ rad/s})\sin[(2.1 \text{ rad/s})(0.50 \text{ s})] = -0.73 \text{ m/s}$$
Thus her speed is 0.73 m/s.

14.37. Model: The spring undergoes simple harmonic motion.
Solve: (a) Total energy is $E = \frac{1}{2}kA^2$. When the displacement is $x = \frac{1}{2}A$, the potential energy is
$$U = \frac{1}{2}kx^2 = \frac{1}{2}k\left(\frac{1}{2}A\right)^2 = \frac{1}{4}\left(\frac{1}{2}kA^2\right) = \frac{1}{4}E \Rightarrow K = E - U = \frac{3}{4}E$$
One quarter of the energy is potential and three-quarters is kinetic.
(b) To have $U = \frac{1}{2}E$ requires
$$U = \frac{1}{2}kx^2 = \frac{1}{2}E = \frac{1}{2}\left(\frac{1}{2}kA^2\right) \Rightarrow x = \frac{A}{\sqrt{2}}$$

14.39. Model: The ball attached to a spring is in simple harmonic motion.
Solve: (a) Let $t = 0$ s be the instant when $x_0 = -5.0$ cm and $v_0 = 20$ cm/s. The oscillation frequency is
$$\omega = \sqrt{\frac{k}{m}} = \sqrt{\frac{2.5 \text{ N/m}}{0.100 \text{ kg}}} = 5.0 \text{ rad/s}$$
Solving Equation 14.26 for the amplitude gives
$$A = \sqrt{x_0^2 + \left(\frac{v_0}{\omega}\right)^2} = \sqrt{(-5.0 \text{ cm})^2 + \left(\frac{20 \text{ cm/s}}{5.0 \text{ rad/s}}\right)^2} = 6.4 \text{ cm}$$
(b) The maximum acceleration is $a_{max} = \omega^2 A = 160 \text{ cm/s}^2$.
(c) For an oscillator, the acceleration is most positive ($a = a_{max}$) when the displacement is most negative ($x = -x_{max} = -A$). So the acceleration is maximum when $x = -6.4$ cm.
(d) We can use the conservation of energy between $x_0 = -5.0$ cm and $x_1 = 3.0$ cm:
$$\frac{1}{2}mv_0^2 + \frac{1}{2}kx_0^2 = \frac{1}{2}mv_1^2 + \frac{1}{2}kx_1^2 \Rightarrow v_1 = \sqrt{v_0^2 + \frac{k}{m}(x_0^2 - x_1^2)} = 0.283 \text{ m/s}$$
The speed is 28 cm/s. Because k is known in SI units of N/m, the energy calculation *must* be done using SI units of m, m/s, and kg.

14.43. Model: The block attached to the spring is in simple harmonic motion.
Solve: (a) The frequency is
$$f = \frac{1}{2\pi}\sqrt{\frac{k}{m}} = \frac{1}{2\pi}\sqrt{\frac{2000 \text{ N/m}}{5.0 \text{ kg}}} = 3.183 \text{ Hz}$$
The frequency is 3.2 Hz.
(b) From energy conservation,
$$A = \sqrt{x_0^2 + \left(\frac{v_0}{\omega}\right)^2} = \sqrt{(0.050 \text{ m})^2 + \left(\frac{1.0 \text{ m/s}}{2\pi \cdot 3.183 \text{ Hz}}\right)^2} = 0.0707 \text{ m}$$
The amplitude is 7.1 cm.
(c) The total mechanical energy is
$$E = \frac{1}{2}kA^2 = \frac{1}{2}(2000 \text{ N/m})(0.0707 \text{ m})^2 = 5.0 \text{ J}$$

14.47. Model: The spring is ideal, so the apples undergo simple harmonic motion.
Solve: The spring constant of the scale can be found by considering how far the pan goes down when the apples are added.

$$\Delta L = \frac{mg}{k} \Rightarrow k = \frac{mg}{\Delta L} = \frac{20\ \text{N}}{0.090\ \text{m}} = 222\ \text{N/m}$$

The frequency of oscillation is

$$f = \frac{1}{2\pi}\sqrt{\frac{k}{m}} = \frac{1}{2\pi}\sqrt{\frac{222\ \text{N/m}}{(20\ \text{N}/9.8\ \text{m/s}^2)}} = 1.66\ \text{Hz} \approx 1.7\ \text{Hz}$$

Assess: An oscillation of fewer than twice per second is reasonable.

14.51. Model: Assume that the swinging lamp makes a small angle with the vertical so that there is simple harmonic motion.
Visualize:

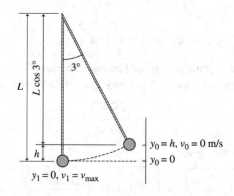

Solve: **(a)** Using the formula for the period of a pendulum,

$$T = 2\pi\sqrt{\frac{L}{g}} \Rightarrow L = g\left(\frac{T}{2\pi}\right)^2 = (9.8\ \text{m/s}^2)\left(\frac{5.5\ \text{s}}{2\pi}\right)^2 = 7.5\ \text{m}$$

(b) The conservation of mechanical energy equation $K_0 + U_{g0} = K_1 + U_{g1}$ for the swinging lamp is

$$\tfrac{1}{2}mv_0^2 + mgy_0 = \tfrac{1}{2}mv_1^2 + mgy_1 \Rightarrow 0\ \text{J} + mgh = \tfrac{1}{2}mv_{max}^2 + 0\ \text{J}$$

$$\Rightarrow v_{max} = \sqrt{2gh} = \sqrt{2g(L - L\cos 3°)}$$

$$= \sqrt{2(9.8\ \text{m/s}^2)(7.5\ \text{m})(1 - \cos 3°)} = 0.45\ \text{m/s}$$

14.53. Model: Assume the orangutan is a simple small-angle pendulum.
Visualize: One swing of the arm would be half a period of the oscillatory motion. The horizontal distance traveled in that time would be $2(0.90\ \text{m})\sin(20°) = 0.616\ \text{m}$ from analysis of a right triangle.
Solve:

$$\frac{T}{2} = \frac{2\pi}{2}\sqrt{\frac{L}{g}} = \frac{2\pi}{2}\sqrt{\frac{0.90\ \text{m}}{9.8\ \text{m/s}^2}} = 0.952\ \text{s}$$

$$\text{speed} = \frac{\text{dist}}{\text{time}} = \frac{0.616\ \text{m}}{0.952\ \text{s}} = 0.65\ \text{m/s}$$

Assess: This isn't very fast, but isn't out of the reasonable range.

14.57. Model: Model the sphere as a small-angle physical pendulum without damping.
Visualize: Figure 14.22 in the chapter shows that l is the distance from the pivot to the center of mass; in this case $l = R$.

Solve: We first need to compute the moment of inertia of a sphere about an axis tangent to the edge. Use the parallel-axis theorem to do so:

$$I = I_{c.m.} + Md^2 = \frac{2}{5}MR^2 + MR^2 = \frac{7}{5}MR^2$$

For the physical pendulum

$$f = \frac{1}{2\pi}\sqrt{\frac{Mgl}{I}} = \frac{1}{2\pi}\sqrt{\frac{MgR}{\frac{7}{5}MR^2}} = \frac{1}{2\pi}\sqrt{\frac{5}{7}\frac{g}{R}}$$

Assess: The frequency is less than for a simple pendulum of length R as we expect.

14.59. Model: Treat the lower leg as a physical pendulum.

Visualize: We can determine the moment of inertia by combining $T = 2\pi\sqrt{I/mgL}$ and $T = 1/f$.

Solve: Combining the above expressions and solving for the moment of inertia we obtain

$$I = mgL/(2\pi f)^2 = (5.0\text{ kg})(9.80\text{ m/s}^2)(0.18\text{ m})/[2\pi(1.6\text{ Hz})]^2 = 8.7\times10^{-2}\text{ kg}\cdot\text{m}^2$$

Assess: NASA determines the moment of inertia of the shuttle in a similar manner. It is suspended from a heavy cable, allowed to oscillate about its vertical axis of symmetry with a very small amplitude, and from the period of oscillation one may determine the moment of inertia. This arrangement is called a torsion pendulum.

14.61. Model: The block attached to the spring is oscillating in simple harmonic motion.
Solve: (a) Because the frequency of an object in simple harmonic motion is independent of the amplitude and/or the maximum velocity, the new frequency is equal to the old frequency of 2.0 Hz.
(b) The speed v_0 of the block just before it is given a blow can be obtained by using the conservation of mechanical energy equation as follows:

$$\tfrac{1}{2}kA^2 = \tfrac{1}{2}mv_{max}^2 = \tfrac{1}{2}mv_0^2$$

$$\Rightarrow v_0 = \sqrt{\frac{k}{m}}A = \omega A = (2\pi f)A = (2\pi)(2.0\text{ Hz})(0.02\text{ m}) = 0.25\text{ m/s}$$

The blow to the block provides an impulse that changes the velocity of the block:

$$J_x = F_x\Delta t = \Delta p = mv_f - mv_0$$

$$(-20\text{ N})(1.0\times10^{-3}\text{ s}) = (0.200\text{ kg})v_f - (0.200\text{ kg})(0.25\text{ m/s}) \Rightarrow v_f = 0.150\text{ m/s}$$

Since v_f is the new maximum velocity of the block at the equilibrium position, it is equal to $A\omega$. Thus,

$$A = \frac{0.150\text{ m/s}}{\omega} = \frac{0.150\text{ m/s}}{2\pi(2.0\text{ Hz})} = 0.012\text{ m} = 1.19\text{ cm} \approx 1.2\text{ cm}$$

Assess: Because v_f is positive, the block continues to move to the right even after the blow.

14.63. Solve: The potential energy curve of a simple harmonic oscillator is described by $U = \frac{1}{2}k(\Delta x)^2$, where $\Delta x = x - x_0$ is the displacement from equilibrium. From the graph, we see that the equilibrium bond length is $x_0 = 0.13$ nm. We can find the bond's spring constant by reading the value of the potential energy U at a displacement Δx and using the potential energy formula to calculate k.

x (nm)	Δx (nm)	U (J)	k (N/m)
0.11	0.02	0.8×10^{-19} J	400
0.10	0.03	1.9×10^{-19} J	422
0.09	0.04	3.4×10^{-19} J	425

The three values of k are all very similar, as they should be, with an average value of 416 N/m. Knowing the spring constant, we can now calculate the oscillation frequency of a hydrogen atom on this "spring" to be

$$f = \frac{1}{2\pi}\sqrt{\frac{k}{m}} = \frac{1}{2\pi}\sqrt{\frac{416 \text{ N/m}}{1.67 \times 10^{-27} \text{ kg}}} = 7.9 \times 10^{13} \text{ Hz}$$

14.67. Model: The doll's head is in simple harmonic motion and is damped.
Solve: (a) The oscillation frequency is

$$f = \frac{1}{2\pi}\sqrt{\frac{k}{m}} \Rightarrow k = m(2\pi f)^2 = (0.015 \text{ kg})(2\pi)^2(4.0 \text{ Hz})^2 = 9.475 \text{ N/m}$$

The spring constant is 9.5 N/m.
(b) Using $A(t) = A_0 e^{-bt/2m}$, we get

$$(0.5 \text{ cm}) = (2.0 \text{ cm})e^{-b(4.0 \text{ s})/(2 \times 0.015 \text{ kg})} \Rightarrow 0.25 = e^{-(133.3 \text{ s/kg})b}$$
$$\Rightarrow -(133.33 \text{ s/kg})b = \ln 0.25 \Rightarrow b = 0.0104 \text{ kg/s} \approx 0.010 \text{ kg/s}$$

14.69. Model: The vertical oscillations are damped and follow simple harmonic motion.
Solve: The position of the ball is given by $x(t) = Ae^{-(t/2\tau)}\cos(\omega t + \phi_0)$. The amplitude $A(t) = Ae^{-(t/2\tau)}$ is a function of time. The angular frequency is

$$\omega = \sqrt{\frac{k}{m}} = \sqrt{\frac{(15.0 \text{ N/m})}{0.500 \text{ kg}}} = 5.477 \text{ rad/s} \Rightarrow T = \frac{2\pi}{\omega} = 1.147 \text{ s}$$

Because the ball's amplitude decreases to 3.0 cm from 6.0 cm after 30 oscillations, that is, after $30 \times 1.147 \text{ s} = 34.41 \text{ s}$, we have

$$3.0 \text{ cm} = (6.0 \text{ cm})e^{-(34.414 \text{ s}/2\tau)} \Rightarrow 0.50 = e^{-(34.41 \text{ s}/2\tau)} \Rightarrow \ln(0.50) = \frac{-34.41 \text{ s}}{2\tau} \Rightarrow \tau = 25 \text{ s}$$

14.73. Model: The two springs obey Hooke's law.
Visualize:

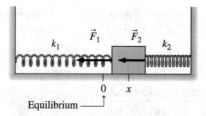

$$\text{Equilibrium}$$

Solve: There are *two* restoring forces on the block. If the block's displacement x is positive, *both* restoring forces—one pushing, the other pulling—are directed to the left and have negative values:

$$(F_{\text{net}})_x = (F_{\text{sp 1}})_x + (F_{\text{sp 2}})_x = -k_1 x - k_2 x = -(k_1 + k_2)x = -k_{\text{eff}} x$$

where $k_{\text{eff}} = k_1 + k_2$ is the effective spring constant. This means the oscillatory motion of the block under the influence of the two springs will be the *same* as if the block were attached to a single spring with spring constant k_{eff}. The frequency of the blocks, therefore, is

$$f = \frac{1}{2\pi}\sqrt{\frac{k_{\text{eff}}}{m}} = \frac{1}{2\pi}\sqrt{\frac{k_1 + k_2}{m}} = \sqrt{\frac{k_1}{4\pi^2 m} + \frac{k_2}{4\pi^2 m}} = \sqrt{f_1^2 + f_2^2}$$

15

FLUIDS AND ELASTICITY

Exercises and Problems

Section 15.1 Fluids

15.1. Solve: The volume of the liquid is

$$\rho = \frac{m}{V} \quad \Rightarrow \quad V = \frac{m}{\rho} = \frac{0.055 \text{ kg}}{1100 \text{ kg/m}^3}\left(\frac{10^6 \text{ mL}}{\text{m}^3}\right) = 50 \text{ mL}$$

Assess: The liquid's density slightly higher than that of water (1000 kg/m^3), so it is reasonable that it requires slightly less than 55 mL to get a mass of 55 g.

15.3. Model: The density of water is 1000 kg/m^3.

Visualize:

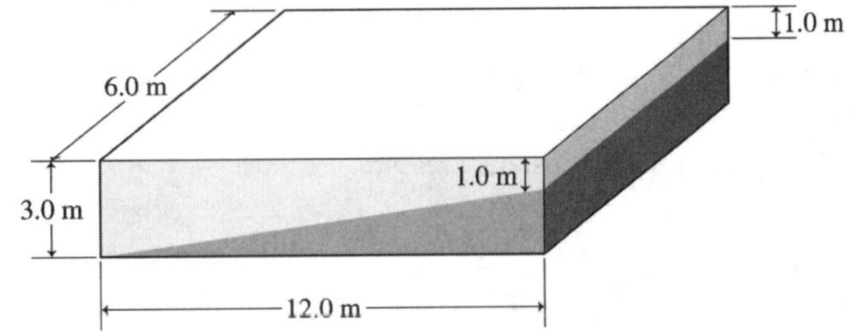

Solve: Volume of water in the swimming pool is

$$V = 6.0 \text{ m} \times 12 \text{ m} \times 3.0 \text{ m} - \tfrac{1}{2}(6.0 \text{ m} \times 12 \text{ m} \times 2.0 \text{ m}) = 144 \text{ m}^3$$

The mass of water in the swimming pool is

$$m = \rho V = (1000 \text{ kg/m}^3)(144 \text{ m}^3) = 1.4 \times 10^5 \text{ kg}$$

Section 15.2 Pressure

15.9. Model: The density of seawater is $\rho_{\text{seawater}} = 1030 \text{ kg/m}^2$.

Visualize:

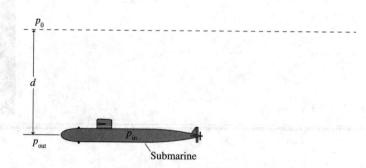

Submarine

Solve: The pressure outside the submarine's window is $p_{out} = p_0 + \rho_{seawater}gd$, where d is the maximum safe depth for the window to withstand a force F. This force is $F/A = p_{out} - p_{in}$, where A is the area of the window. With $p_{in} = p_0$, we simplify the pressure equation to

$$p_{out} - p_0 = \frac{F}{A} = \rho_{seawater}gd \quad \Rightarrow \quad d = \frac{F}{A\rho_{seawater}g}$$

$$d = \frac{1.0 \times 10^6 \text{ N}}{\pi(0.10 \text{ m})^2(1030 \text{ kg/m}^2)(9.8 \text{ m/s}^2)} = 3.2 \text{ km}$$

Assess: A force of 1.0×10^6 N corresponds to a pressure of

$$\rho = \frac{F}{A} = \frac{1.0 \times 10^6 \text{ N}}{\pi(0.10 \text{ m})^2} = 314 \text{ atm}$$

A depth of 3 km is therefore reasonable.

Section 15.3 Measuring and Using Pressure

15.11. Model: The density of water is $\rho = 1000$ kg/m^3.
Visualize: Please refer to Figure 15.16.
Solve: From the figure and the equation for hydrostatic pressure, we have

$$p_0 + \rho gh = p_{atmos}$$

Using $p_0 = 0$ atm, and $p_{atmos} = 1.013 \times 10^5$ Pa, we get

$$0 \text{ Pa} + (1000 \text{ kg/m}^3)(9.81 \text{ m/s}^2)h = 1.013 \times 10^5 \text{ Pa} \quad \Rightarrow \quad h = 10.3 \text{ m}$$

Assess: This large value of h is due to water having a much smaller density than mercury.

15.13. Model: Assume that the vacuum cleaner can create zero pressure.
Visualize:

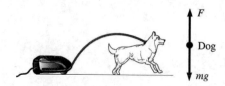

Solve: The gravitational force on the dog is balanced by the force resulting from the pressure difference between the atmosphere and the vacuum $(p_{\text{hose}} = 0)$ in the hose. The force applied by the hose is

$$F = (p_{\text{atmos}} - p_{\text{hose}})A = p_{\text{atmos}}A = mg$$

$$A = \frac{(10 \text{ kg})(9.8 \text{ m/s}^2)}{1.013\times10^5 \text{ Pa}} = 9.7\times10^{-4} \text{ m}^2$$

Since $A = \pi(d/2)^2$, the diameter of the hose is $d = 2\sqrt{A/\pi} = 0.035 \text{ m} = 3.5 \text{ cm}$.

Section 15.4 Buoyancy

15.17. Model: The buoyant force on the rock is given by Archimedes' principle.
Visualize:

Solve: Because the rock is in static equilibrium, Newton's first law gives

$$F_{\text{net}} = T + F_{\text{B}} - (F_{\text{G}})_{\text{rock}} = 0 \text{ N}$$

$$T = \rho_{\text{rock}}V_{\text{rock}}g - \rho_{\text{water}}\left(\frac{1}{2}V_{\text{rock}}\right)g = \left(\rho_{\text{rock}} - \frac{1}{2}\rho_{\text{water}}\right)V_{\text{rock}}g = \left(\rho_{\text{rock}} - \frac{1}{2}\rho_{\text{water}}\right)\left(\frac{m_{\text{rock}}g}{\rho_{\text{rock}}}\right) = \left(1 - \frac{\rho_{\text{water}}}{2\rho_{\text{rock}}}\right)m_{\text{rock}}g$$

Using $\rho_{\text{rock}} = 4800 \text{ kg/m}^3$ and $m_{\text{rock}} = 5.0 \text{ kg}$, we get $T = 44 \text{ N}$.

15.19. Model: The buoyant force on the steel cylinder is given by Archimedes' principle.
Visualize:

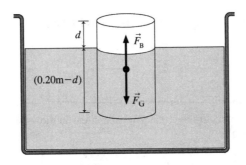

The length of the cylinder above the surface of mercury is d.
Solve: The cylinder is in static equilibrium with $F_{\text{B}} = F_{\text{G}}$. Thus

$$F_{\text{B}} = \rho_{\text{Hg}}V_{\text{Hg}}g = F_{\text{G}} = mg = \rho_{\text{cyl}}V_{\text{cyl}}g \implies \rho_{\text{Hg}}V_{\text{Hg}} = \rho_{\text{cyl}}V_{\text{cyl}} \implies \rho_{\text{Hg}}A(0.20 \text{ m} - d) = \rho_{\text{cyl}}A(0.20 \text{ m})$$

$$d = 0.20 \text{ m} - \frac{\rho_{\text{cyl}}}{\rho_{\text{Hg}}}(0.20 \text{ m}) = (0.20 \text{ m})\left(1 - \frac{7900 \text{ kg/m}^3}{13{,}600 \text{ kg/m}^3}\right) = 0.084 \text{ m} = 8.4 \text{ cm}$$

That is, the length of the cylinder above the surface of the mercury is 8.4 cm.

15.21. Model: The buoyant force on the sphere is given by Archimedes' principle.
Visualize:

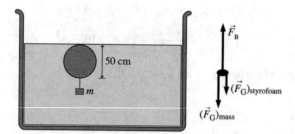

Solve: For the Styrofoam sphere and the mass not to sink, the sphere must be completely submerged and the buoyant force F_B must be equal to the sum of the gravitational force on the Styrofoam sphere and the attached mass. Neglecting the volume of the hanging mass, the volume of displaced water equals the volume of the sphere, so

$$F_B = \rho_{water} V_{water} g = (1000 \text{ kg/m}^3)\left(\tfrac{4\pi}{3}\right)(0.25 \text{ m})^3 (9.8 \text{ m/s}^2) = 641.4 \text{ N}$$

$$(F_G)_{Styrofoam} = \rho_{Styrofoam} V_{Styrofoam} g = (150 \text{ kg/m}^3)\left[\tfrac{4}{3}\pi(0.25 \text{ m})^3\right](9.8 \text{ m/s}^2) = 96.2 \text{ N}$$

Because $(F_G)_{Styrofoam} + mg = F_B$,

$$m = \frac{F_B - (F_G)_{Styrofoam}}{g} = \frac{641.4 \text{ N} - 96.2 \text{ N}}{9.8 \text{ m/s}^2} = 55.6 \text{ kg}$$

To two significant figures, the mass is 56 kg.

Section 15.5 Fluid Dynamics

15.25. Model: Treat the oil as an ideal fluid obeying Bernoulli's equation. Consider the path connecting point 1 in the lower pipe with point 2 in the upper pipe a streamline.
Visualize: Please refer to Figure EX15.25.
Solve: Bernoulli's equation is

$$p_2 + \tfrac{1}{2}\rho v_2^2 + \rho g y_2 = p_1 + \tfrac{1}{2}\rho v_1^2 + \rho g y_1 \quad \Rightarrow \quad p_2 = p_1 + \tfrac{1}{2}\rho(v_1^2 - v_2^2) + \rho g(y_1 - y_2)$$

Using $p_1 = 200 \text{ kPa} = 2.0 \times 10^5 \text{ Pa}$, $\rho = 900 \text{ kg/m}^3$, $y_2 - y_1 = 10.0 \text{ m}$, $v_1 = 2.0 \text{ m/s}$, and $v_2 = 3.0 \text{ m/s}$, we get $p_2 = 1.096 \times 10^5 \text{ Pa} = 110 \text{ kPa}$.

Section 15.6 Elasticity

15.27. Model: The dangling mountain climber creates tensile stress in the rope.
Solve: Young's modulus for the rope is

$$Y = \frac{F/A}{\Delta L/L} = \frac{\text{stress}}{\text{strain}}$$

The tensile stress is

$$\frac{(70 \text{ kg})(9.8 \text{ m/s}^2)}{\pi(0.0050 \text{ m})^2} = 8.734 \times 10^6 \text{ Pa}$$

and the strain is $0.080 \text{ m}/50 \text{ m} = 0.00160$. Dividing the two quantities yields $Y = 5.5 \times 10^9 \text{ N/m}^2$.

15.31. Solve: The pressure p at depth d in a fluid is $p = p_0 + \rho gd$. Using 1.29 kg/m^3 for the density of air,

$$p_{\text{bottom}} = p_{\text{top}} + \rho_{\text{air}}gd \quad \Rightarrow \quad p_{\text{bottom}} - p_{\text{top}} = (1.29 \text{ kg/m}^3)(9.8 \text{ m/s}^2)(16 \text{ m}) = 202 \text{ Pa} = 1.99 \times 10^{-3} \text{ atm}$$

Assuming $p_{\text{bottom}} = 1$ atm,

$$\frac{p_{\text{bottom}} - p_{\text{top}}}{p_{\text{bottom}}} = \frac{1.99 \times 10^{-3} \text{ atm}}{1 \text{ atm}} = 0.20\%$$

15.35. Model: The tire flattens until the pressure force against the ground balances the upward normal force of the ground on the tire.
Solve: The area of the tire in contact with the road is $A = (0.15 \text{ m})(0.13 \text{ m}) = 0.0195 \text{ m}^2$. The normal force on each tire is

$$n = \frac{F_G}{4} = \frac{(1500 \text{ kg})(9.8 \text{ m/s}^2)}{4} = 3675 \text{ N}$$

Thus, the pressure inside each tire is

$$p_{\text{inside}} = \frac{n}{A} = \frac{3675 \text{ N}}{0.0195 \text{ m}^2} = 188,500 \text{ Pa} = 1.86 \text{ atm} \times \frac{14.7 \text{ psi}}{1 \text{ atm}} = 27 \text{ psi}$$

15.37. Solve: The fact that atmospheric pressure at sea level is 101.3 kPa $= 101,300$ N/m^2 means that the weight of the atmosphere over each square meter of surface is $101,300$ N. Thus the mass of air over each square meter is $m = (101,300 \text{ N})/g = (101,300 \text{ N})/(9.80 \text{ m/s}^2) = 10,340$ kg per m^2. Multiplying by the earth's surface area will give the total mass. Using $R_e = 6.27 \times 10^6$ m for the earth's radius, the total mass of the atmosphere is

$$M_{\text{air}} = A_{\text{earth}}m = (4\pi R_e^2)m = 4\pi(6.37 \times 10^6 \text{ m})^2(10,340 \text{ kg/m}^2) = 5.27 \times 10^{18} \text{ kg}$$

15.39. Model: Oil is incompressible and has a density 900 kg/m^3.
Visualize: Please refer to Figure P15.39.
Solve: (a) The pressure at point A, which is 0.50 m below the open oil surface, is

$$p_A = p_0 + \rho_{\text{oil}}g(1.00 \text{ m} - 0.50 \text{ m}) = 101,300 \text{ Pa} + (900 \text{ kg/m}^3)(9.8 \text{ m/s}^2)(0.50 \text{ m}) = 106 \text{ kPa}$$

(b) The pressure difference between A and B is

$$p_B - p_A = (p_0 + \rho gd_B) - (p_0 + \rho gd_A) = \rho g(d_B - d_A) = (900 \text{ kg/m}^3)(9.8 \text{ m/s}^2)(0.50 \text{ m}) = 4.4 \text{ kPa}$$

Pressure depends only on depth, and C is the same depth as B. Thus $p_C - p_A = 4.4$ kPa also, even though C isn't directly under A.

15.43. Model: Glycerin and ethyl alcohol are incompressible and do not mix.
Visualize:

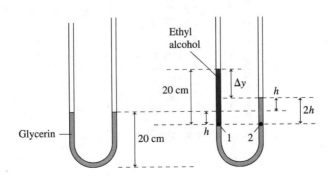

Solve: The alcohol in the left arm floats on top of the denser glycerin and presses the glycerin down distance h from its initial level. This causes the glycerin to rise distance h in the right arm. Points 1 and 2 are level with each other *and* the fluids are in static equilibrium, so the pressures at these two points must be equal:

$$p_1 = p_2 \quad \Rightarrow \quad p_0 + \rho_{eth}gd_{eth} = p_0 + \rho_{gly}gd_{gly} \quad \Rightarrow \quad \rho_{eth}g(20 \text{ cm}) = \rho_{gly}g(2h)$$

$$h = \frac{1}{2}\frac{\rho_{eth}}{\rho_{gly}}(20 \text{ cm}) = \left(\frac{1}{2}\right)\left(\frac{790 \text{ kg/m}^3}{1260 \text{ kg/m}^3}\right)(20 \text{ cm}) = 6.27 \text{ cm}$$

You can see from the figure that the difference between the top surfaces of the fluids is

$$\Delta y = 20 \text{ cm} - 2h = 20 \text{ cm} - 2(6.27 \text{ cm}) = 7.46 \text{ cm} \approx 7.5 \text{ cm}$$

15.47. Model: Ignore the change in water density over lengths comparable to the size of the fish.
Visualize:

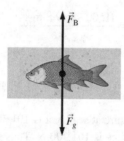

Solve: The buoyant force is the force of gravity on the volume of water displaced by the fish times, or

$$F_B = \rho_w(V_F + V_b)g$$

where the subscripts w, F, and b indicate water, fish, and bladder, respectively. The force due to gravity on the fish is

$$F_G = \rho_F V_F g + \rho_{air} V_b g$$

For neutral buoyancy, the two forces must have equal magnitude. Equating them and solving for the V_b gives

$$\rho_w(V_F + V_b)g = \rho_F V_F g + \rho_{air} V_b g$$

$$V_b = \frac{\rho_w - \rho_F}{\rho_{air} - \rho_w}V_F = \frac{1000 \text{ kg/m}^3 - 1080 \text{ kg/m}^3}{1.19 \text{ kg/m}^3 - 1000 \text{ kg/m}^3}V_F = 0.0801V_F$$

so the fish needs to increase its volume by 8.01%.

15.51. Model: The buoyant force on the cylinder is given by Archimedes' principle.
Visualize:

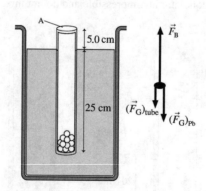

Solve: The tube is in static equilibrium, so

$$\sum F_y = F_B - (F_G)_{tube} - (F_G)_{Pb} = 0 \text{ N} \quad \Rightarrow \quad \rho_{liquid} A (0.25 \text{ m}) g = (0.030 \text{ kg}) g + (0.250 \text{ kg}) g$$

$$\rho_{liquid} = \frac{(0.280 \text{ kg})}{A(0.25 \text{ m})} = \frac{(0.280 \text{ kg})}{\pi (0.020 \text{ m})^2 (0.25 \text{ m})} = 8.9 \times 10^2 \text{ kg/m}^3$$

Assess: This is a reasonable value for a liquid.

15.53. Model: The buoyant force is determined by Archimedes' principle. The spring is ideal.
Visualize:

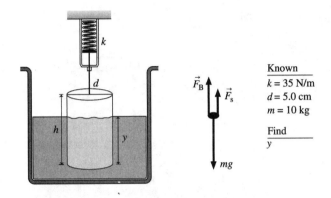

Known
$k = 35 \text{ N/m}$
$d = 5.0 \text{ cm}$
$m = 10 \text{ kg}$

Find
y

Solve: The spring is stretched by the same amount that the cylinder is submerged. The buoyant force and spring force balance the gravitational force on the cylinder.

$$\sum F_y = F_B + F_S - mg = 0 \text{ N}$$

$$\rho_w Ayg + ky = mg$$

$$y = \frac{mg}{\rho_w Ag + k} = \frac{(1.0 \text{ kg})(9.8 \text{ m/s}^2)}{(1000 \text{ kg/m}^3)\pi(0.025 \text{ m})^2(9.8 \text{ m/s}^2) + 35 \text{ N/m}}$$

$$= 0.181 \text{ m} = 18 \text{ cm}$$

Assess: This is difficult to assess because we don't know the height h of the cylinder and can't calculate it without the density of the metal material.

15.55. Model: The buoyant force on the can is given by Archimedes' principle.
Visualize:

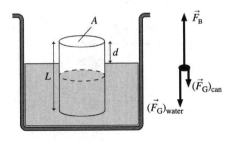

The length of the can above the water level is d, the length of the can is L, and the cross-sectional area of the can is A.
Solve: The can is in static equilibrium, so

$$\sum F_y = F_B - (F_G)_{can} - (F_G)_{water} = 0 \text{ N} \tfrac{1}{2} \rho_{water} A (L-d) g = (0.020 \text{ kg}) g + m_{water} g$$

The mass of the water in the can is

$$m_{\text{water}} = \rho_{\text{water}}\left(\frac{V_{\text{can}}}{2}\right) = (1000 \text{ kg/m}^3)\frac{355\times10^{-6} \text{ m}^3}{2} = 0.1775 \text{ kg}$$

$$\rho_{\text{water}}A(L-d) = 0.020 \text{ kg} + 0.1775 \text{ kg} = 0.1975 \text{ kg} \implies d - L = -\frac{0.1975 \text{ kg}}{\rho_{\text{water}}A} = 0.0654 \text{ m}$$

Because $V_{\text{can}} = \pi(0.031 \text{ m})^2 L = 355\times10^{-6} \text{ m}^3$, $L = 0.1176 \text{ m}$. Using this value of L, we get $d = 0.0522 \text{ m} \approx 5.2 \text{ cm}$. **Assess:** $d/L = 5.22 \text{ cm}/11.76 \text{ cm} = 0.444$, thus 44.4% of the length of the can is above the water surface. This is reasonable.

15.57. Model: The two pipes are identical.
Visualize:

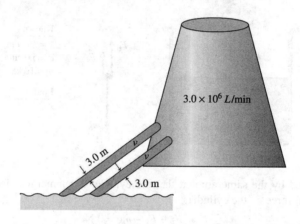

Solve: The water speed is the same in both pipes. The flow rate is

$$Q = 3.0\times10^6 \text{ L/min} = 2(vA)$$

$$v = \frac{Q}{2A} = \frac{(3.0\times10^6)(10^{-3})\left(\dfrac{1}{60}\right) \text{ m}^3/\text{s}}{2\pi(1.5 \text{ m})^2} = 3.5 \text{ m/s}$$

15.61. Model: Treat the water as an ideal fluid obeying Bernoulli's equation. A streamline begins at the faucet and continues down the stream.
Visualize:

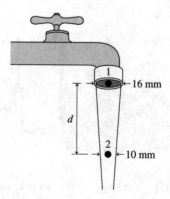

The pressure at point 1 is p_1 and the pressure at point 2 is p_2. Both p_1 and p_2 are atmospheric pressure. The velocity and the area at point 1 are v_1 and A_1 and they are v_2 and A_2 at point 2. Let d be the distance of point 2 below point 1.

Solve: The flow rate is

$$Q = v_1 A = \frac{2.0 \times 1000 \times 10^{-6} \text{ m}^3}{10 \text{ s}} = 2.0 \times 10^{-4} \text{ m}^3/\text{s} \quad \Rightarrow \quad v_1 = \frac{2.0 \times 10^{-4} \text{ m}^3/\text{s}}{\pi(0.0080 \text{ m})^2} = 1.0 \text{ m/s}$$

Bernoulli's equation at points 1 and 2 is

$$p_1 + \tfrac{1}{2}\rho v_1^2 + \rho g y_1 = p_2 + \tfrac{1}{2}\rho v_2^2 + \rho g y_2 \quad \Rightarrow \quad \rho g d = \tfrac{1}{2}\rho(v_2^2 - v_1^2)$$

From the continuity equation,

$$v_1 A_1 = v_2 A_2 \quad \Rightarrow \quad (1.0 \text{ m/s})\pi(8.0 \times 10^{-3} \text{ m})^2 = v_2\pi(5.0 \times 10^{-3} \text{ m})^2 \quad \Rightarrow \quad v_2 = 2.56 \text{ m/s}$$

Going back to Bernoulli's equation, we have

$$g d = \tfrac{1}{2}[(2.56 \text{ m/s})^2 - (1.0 \text{ m/s})^2] \quad \Rightarrow \quad d = 0.283 \text{ m} \approx 28 \text{ cm}$$

15.63. Model: The ideal fluid obeys Bernoulli's equation.
Visualize: Please refer to Figure P15.63. There is a streamline connecting point 1 in the wider pipe on the left with point 2 in the narrower pipe on the right. The air speeds at points 1 and 2 are v_1 and v_2 and the cross-sectional area of the pipes at these points are A_1 and A_2. Points 1 and 2 are at the same height, so $y_1 = y_2$.
Solve: The volume flow rate is $Q = A_1 v_1 = A_2 v_2 = 1200 \times 10^{-6} \text{ m}^3/\text{s}$. Thus

$$v_2 = \frac{1200 \times 10^{-6} \text{ m}^3/\text{s}}{\pi(0.0020 \text{ m})^2} = 95.49 \text{ m/s} \qquad v_1 = \frac{1200 \times 10^{-6} \text{ m}^3/\text{s}}{\pi(0.010 \text{ m})^2} = 3.82 \text{ m/s}$$

Now we can use Bernoulli's equation to connect points 1 and 2:

$$p_1 + \tfrac{1}{2}\rho v_1^2 + \rho g y_1 = p_2 + \tfrac{1}{2}\rho v_2^2 + \rho g y_2$$

$$p_1 - p_2 = \tfrac{1}{2}\rho(v_2^2 - v_1^2) + \rho g(y_2 - y_1) = \tfrac{1}{2}(1.28 \text{ kg/m}^3)[(95.49 \text{ m/s})^2 - (3.82 \text{ m/s})^2] + 0 \text{ Pa} = 5.83 \text{ kPa}$$

Because the pressure above the mercury surface in the right tube is p_2 and in the left tube is p_1, the difference in the pressures p_1 and p_2 is $\rho_{Hg} g h$. That is,

$$p_1 - p_2 = 5.83 \text{ kPa} = \rho_{Hg} g h \quad \Rightarrow \quad h = \frac{5.83 \times 10^3 \text{ Pa}}{(13,600 \text{ kg/m}^3)(9.8 \text{ m/s}^2)} = 4.4 \text{ cm}$$

15.67. Model: The aquarium creates tensile stress.
Solve: Weight of the aquarium is

$$F_G = mg = \rho_{water} V g = (1000 \text{ kg/m}^3)(10 \text{ m}^3)(9.8 \text{ m/s}^2) = 9.8 \times 10^4 \text{ N}$$

where we have used the conversion $1 \text{ L} = 10^{-3} \text{ m}^3$. The weight supported by each wood post is $\tfrac{1}{4}(9.8 \times 10^4 \text{ N}) = 2.45 \times 10^4 \text{ N}$. The cross-sectional area of each post is $A = (0.040 \text{ m})^2 = 1.6 \times 10^{-3} \text{ m}^2$. Young's modulus for the wood is

$$Y = 1 \times 10^{10} \text{ N/m}^2 = \frac{F/A}{\Delta L/L} = \frac{FL}{A\Delta L}$$

$$\Delta L = \frac{FL}{AY} = \frac{(2.45 \times 10^4 \text{ N})(0.80 \text{ m})}{(1.6 \times 10^{-3} \text{ m}^2)(1 \times 10^{10} \text{ N/m}^2)} = 1.23 \times 10^{-3} \text{ m} = 1 \text{ mm}$$

Assess: A compression of 1 mm due to a weight of $2.45 \times 10^4 \text{ N}$ is reasonable.

15.69. Model: Pressure applies a volume stress to water in the cylinder.
Solve: The volume strain of water due to the pressure applied is

$$\frac{\Delta V}{V} = -\frac{p}{B} = -\frac{2 \times 10^6 \text{ Pa}}{0.2 \times 10^{10} \text{ Pa}} = -1 \times 10^{-3}$$

$$\Delta V = V' - V = -(1 \times 10^{-3})(1.30 \text{ m}^3) = -1.30 \times 10^{-3} \text{ m}^3 \approx -1 \text{ L}$$

As the safety plug on the top of the cylinder bursts, the water comes back to atmospheric pressure. To the precision of the data (one significant figure), one liter of water comes out.

16

A MACROSCOPIC DESCRIPTION OF MATTER

Exercises and Problems

Section 16.1 Solids, Liquids, and Gases

16.1. Model: Recall the density of water is 1000 kg/m^3.

Solve: The mass of gold $m_{Au} = \rho_{Au} V_{Au} = (19,300 \text{ kg/m}^3)(100 \times 10^{-6} \text{ m}^3) = 1.93 \text{ kg}$. For water to have the same mass its volume must be

$$V_{water} = \frac{m_{water}}{\rho_{water}} = \frac{1.93 \text{ kg}}{1000 \text{ kg/m}^3} = 0.00193 \text{ m}^3 \approx 1900 \text{ cm}^3$$

Assess: Since the lead is 19.3 times as dense we expect the water to take 19.3 times the volume.

16.3. Model: The volume of a hollow sphere is

$$V = \frac{4\pi}{3}(r_{out}^3 - r_{in}^3)$$

Solve: We are given $m = 0.690 \text{ kg}$, $r_{out} = 0.050 \text{ m}$, and we know that for aluminum $\rho = 2700 \text{ kg/m}^3$. Solve the above equation for r_{in}.

$$r_{in} = \sqrt[3]{r_{out}^3 - \frac{V}{\frac{4}{3}\pi}}$$

$$= \sqrt[3]{r_{out}^3 - \frac{m/\rho}{\frac{4}{3}\pi}}$$

$$= \sqrt[3]{0.050 \text{ m}^3 - \frac{0.690 \text{ kg}/2700 \text{ kg/m}^3}{\frac{4}{3}\pi}}$$

$$= 0.040 \text{ m}$$

So the inner diameter is 8.0 cm.
Assess: We are happy that the inner diameter is less than the outer diameter, and in a reasonable range.

Section 16.2 Atoms and Moles

16.9. Solve: The mass of mercury is

$$M = \rho V = (13,600 \text{ kg/m}^3)(10 \text{ cm}^3)\left(\frac{10^{-6} \text{ m}^3}{1 \text{ cm}^3}\right) = 0.136 \text{ kg} = 136 \text{ g}$$

and the number of moles is

$$n = \frac{M}{M_{\text{mol}}} = \frac{0.136 \text{ g}}{201 \text{ g/mol}} = 0.6766 \text{ mol}$$

The mass of aluminum with 0.6766 mol of Al is

$$M = (0.6766 \text{ mol})M_{\text{mol}} = (0.6766 \text{ mol})\left(\frac{27 \text{ g}}{\text{mol}}\right) = 18.27 \text{ g} = 0.01827 \text{ kg}$$

This mass M of aluminum corresponds to a volume of

$$V = \frac{M}{\rho} = \frac{0.01827 \text{ kg}}{2700 \text{ kg/m}^3} = 6.8 \times 10^{-6} \text{ m}^3 = 6.8 \text{ cm}^3$$

Assess: We expected an answer in the same order of magnitude. The size of atoms doesn't vary as much as the density of atoms from element to element.

Section 16.3 Temperature

Section 16.4 Phase Changes

16.11. Solve: The lowest temperature is

$$T_{\text{F}} = \tfrac{9}{5}T_{\text{C}} + 32° \Rightarrow -127°\text{F} = \tfrac{9}{5}T_{\text{C}} + 32° \Rightarrow T_{\text{C}} = -88°\text{C} \Rightarrow T_{\text{k}} = (-88.3 + 273) \text{ K} = 185 \text{ K}$$

In the same way, the highest temperature is

$$136°\text{F} = \tfrac{9}{5}T_{\text{C}} + 32° \Rightarrow T_{\text{C}} = 58°\text{C} = 331 \text{ K}$$

Assess: On the absolute scale the highest recorded temperature is not quite twice the lowest.

16.13. Model: A temperature scale is a linear scale.
Solve: (a) We need a conversion formula for °C to °Z, analogous to the conversion of °C to °F. Since temperature scales are linear, $T_{\text{C}} = aT_{\text{Z}} + b$, where a and b are constants to be determined. We know the boiling point of liquid nitrogen is 0°Z and −196°C. Similarly, the melting point of iron is 1000°Z and 1538°C. Thus

$$-196 = 0a + b$$
$$1538 = 1000a + b$$

From the first, $b = -196°$. Then from the second, $a = (1538 + 196)/1000 = 1734/1000$. Thus the conversion is $T_{\text{C}} = (1734/1000)T_{\text{Z}} - 196°$. Since the boiling point of water is $T_{\text{C}} = 100°\text{C}$, its temperature in °Z is

$$T_{\text{Z}} = \left(\frac{1000}{1734}\right)(100° + 196°) = 171°\text{Z}$$

(b) A temperature $T_z = 500°Z$ is

$$T_{\text{C}} = \left(\frac{1734}{1000}\right)500° - 196° = 671°\text{C} = 944 \text{ K}$$

Section 16.5 Ideal Gases

16.15. Model: Treat the nitrogen gas in the closed cylinder as an ideal gas.
Solve: (a) The density before and after the compression are $\rho_{\text{before}} = m_1/V_1$ and $\rho_{\text{after}} = m_2/V_2$. Noting that $m_1 = m_2$ and $V_2 = \tfrac{1}{2}V_1$,

$$\frac{\rho_{\text{after}}}{\rho_{\text{before}}} = \frac{m}{V_2}\frac{V_1}{m} = 2 \Rightarrow \rho_{\text{after}} = 2\rho_{\text{before}}$$

The mass density has changed by a factor of 2.
(b) The number of atoms in the gas is unchanged, implying that the number of moles in the gas remains the same; hence the number density is unchanged.

16.19. Model: We'll assume that air is an ideal gas so we can use the ideal gas law, $pV = nRT$.

Visualize: We are given $V = 5.0$ L $= 0.0050$ m^3, $p = 1$ atm $= 101.3$ kPa, and $T = 37°C = 310$ K.

Also recall that $R = 8.31$ J/(mol·K) and oxygen makes up 20% of the air.

Solve: Solve Equation 12.12 for n, the number of moles of air.

$$n = \frac{pV}{RT} = \frac{(101.3 \text{ kPa})(0.0050 \text{ m}^3)}{(8.31 \text{ J/(mol·K)})(310 \text{ K})} = 0.20 \text{ mol}$$

Multiply the number of moles of air by 20% to get the number of moles of oxygen: $(0.20 \text{ mol})(0.20) = (0.040 \text{ mol})$
$(6.02 \times 10^{23}$ molecules/mol$) = 2.4 \times 10^{22}$ molecules of oxygen.

Assess: The answer is a small number of moles of oxygen, but a large number of molecules of oxygen.

16.21. Model: Treat the neon gas in the sealed cylinder as an ideal gas.

Solve: The volume of the cylinder is $V = \pi r^2 h = \pi(0.05 \text{ m})^2(0.30 \text{ m}) = 2.356 \times 10^{-3}$ m^3. The gauge pressure of the

gas is $120 \text{ psi} \times \dfrac{1 \text{ atm}}{14.7 \text{ psi}} \times \dfrac{1.013 \times 10^5 \text{ Pa}}{1 \text{ atm}} = 8.269 \times 10^5$ Pa, so the absolute pressure of the gas is 8.269×10^5 Pa +

1.013×10^5 Pa $= 9.282 \times 10^5$ Pa. The temperature of the gas is $T = (273 + 30)$ K $= 303$ K. The number of moles of the
gas in the cylinder is

$$n = \frac{pV}{RT} = \frac{(9.282 \times 10^5 \text{ Pa})(2.356 \times 10^{-3} \text{ m}^3)}{(8.31 \text{ J/mol K})(303 \text{ K})} = 0.869 \text{ mol}$$

The mass of the helium is

$$M = nM_{\text{mol}} = (0.869 \text{ mol})(20 \text{ g/mol}) = 17.37 \text{ g} = 17.37 \times 10^{-3} \text{ kg}$$

The mass density is

$$\rho = \frac{M}{V} = \frac{17.37 \times 10^{-3} \text{ kg}}{2.356 \times 10^{-3} \text{ m}^3} = 7.4 \text{ kg/m}^3$$

Section 16.6 Ideal-Gas Processes

16.25. Model: The gas is assumed to be ideal, and since the container is rigid $V_2 = V_1$.

Solve: Convert both temperatures to the Kelvin scale.

$$\frac{p_1 V_1}{T_1} = \frac{p_2 V_2}{T_1} \Rightarrow p_2 = p_1 \frac{T_2}{T_1} = 3 \text{ atm}\left(\frac{253 \text{ K}}{293 \text{ K}}\right) = 2.6 \text{ atm}$$

Assess: On the absolute scale the temperature only went up a little bit, so we expect the pressure to rise a little bit.

16.27. Model: In an isochoric process, the volume of the container stays unchanged. Argon gas in the container is
assumed to be an ideal gas.

Solve: (a) The container has only argon inside with $n = 0.1$ mol, $V_1 = 50$ cm$^3 = 50 \times 10^{-6}$ m^3, and $T_1 = 20°C = 293$ K.
This produces a pressure

$$p_1 = \frac{nRT}{V_1} = \frac{(0.1 \text{ mol})(8.31 \text{ J/mol K})(293 \text{ K})}{50 \times 10^{-6} \text{ m}^3} = 4.87 \times 10^6 \text{ Pa} = 4870 \text{ kPa} \approx 4900 \text{ kPa}$$

An ideal gas process has $p_2 V_2 / T_2 = p_1 V_1 / T_1$. Isochoric heating to a final temperature $T_2 = 300°C = 573$ K has
$V_2 = V_1$, so the final pressure is

$$p_2 = \frac{V_1}{V_2} \frac{T_2}{T_1} p_1 = 1 \times \frac{573}{293} \times 4870 \text{ kPa} = 9520 \text{ kPa} \approx 9500 \text{ kPa}$$

Note that it is essential to express temperatures in kelvins.

(b)

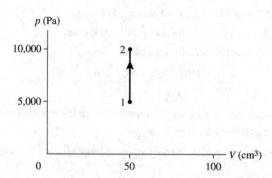

Assess: All isochoric processes will be a straight vertical line on a pV diagram.

16.33. Visualize:

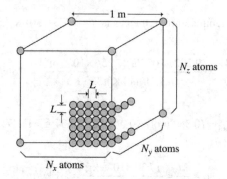

Solve: Suppose we have a $1\,\text{m} \times 1\,\text{m} \times 1\,\text{m}$ block of copper of mass M containing N atoms. The atoms are spaced a distance L apart along all three axes of the cube. There are N_x atoms along the x-edge of the cube, N_y atoms along the y-edge, and N_z atoms along the z-edge. The total number of atoms is $N = N_x N_y N_z$. If L is expressed in meters, then the number of atoms along the x-edge is $N_x = (1\,\text{m})/L$. Thus,

$$N = \frac{1\,\text{m}}{L} \times \frac{1\,\text{m}}{L} \times \frac{1\,\text{m}}{L} = \frac{1\,\text{m}^3}{L^3} \Rightarrow L = \left(\frac{1\,\text{m}^3}{N}\right)^{1/3}$$

This relates the spacing between atoms to the number of atoms in a 1-meter cube. The mass of the large cube of copper is

$$M = \rho_{\text{Cu}} V = (8920\,\text{kg/m}^3)(1\,\text{m}^3) = 8920\,\text{kg}$$

But $M = mN$, where $m = 64\,\text{u} = 64 \times (1.661 \times 10^{-27}\,\text{kg})$ is the mass of an individual copper atom. Thus,

$$N = \frac{M}{m} = \frac{8920\,\text{kg}}{64 \times (1.661 \times 10^{-27}\,\text{kg})} = 8.39 \times 10^{28}\,\text{atoms}$$

$$\Rightarrow L = \left(\frac{1\,\text{m}^3}{8.39 \times 10^{28}}\right)^{1/3} = 2.28 \times 10^{-10}\,\text{m} = 0.228\,\text{nm}$$

Assess: This is a reasonable interatomic spacing in a crystal lattice.

16.35. Model: Assume the density of the liquid water is $\dfrac{1.0\,\text{g}}{1.0\,\text{cm}^3}$.

Visualize: There are 10 protons in each water molecule. 1 mol of water molecules has a mass of 18 g.

Solve:

$$1.0 \text{ L} = 1000 \text{ cm}^3 \left(\frac{1.0 \text{ g}}{1.0 \text{ cm}^3}\right)\left(\frac{1 \text{ mol}}{18 \text{ g}}\right)\left(\frac{6.02\times10^{23} \text{ molecules}}{1 \text{ mol}}\right)\left(\frac{10 \text{ protons}}{1 \text{ molecule}}\right) = 3.3\times10^{26} \text{ protons}$$

Assess: This is an unimaginably large number, but reasonable considering the data.

16.39. Model: Model the oxygen molecules as an ideal gas.

Visualize: Do some preliminary calculations. $n = 100 \text{ mg O}_2 \left(\frac{1 \text{ mol}}{32 \text{ g}}\right) = 0.003125 \text{ mol}$. The area of the cap is $A = \pi r^2 =$

$\pi(0.030 \text{ m})^2 = 0.002827 \text{ m}^2$. The volume of the cylinder is $V = \pi r^2 L = \pi(0.030 \text{ m})^2(0.10 \text{ m}) = 0.0002827 \text{ m}^3$.
The pressure inside the cylinder will be the outside pressure minus the force needed to remove the cap divided by the
area of the cap. $p = 100 \text{ kpa} - \frac{184 \text{ N}}{0.002827 \text{ m}^2} = 34.9\times10^3 \text{ Pa}$.

Solve: Solve the ideal gas law for temperature.

$$T = \frac{pV}{nR} = \frac{(34.9\times10^3 \text{ Pa})(0.0002827 \text{ m}^3)}{(0.003125 \text{ mol})(8.31 \text{ J/mol}\cdot\text{K})} = 380 \text{ K} = 107°\text{C}$$

Assess: The answer is much hotter than room temperature, but the pressure was pretty low.

16.41. Model: Assume the air is pure N_2, with a molar mass of $M_{mol} = 28 \text{ g/mol}$.

Visualize: We will use $pV = nRT$ both before and after. Our intermediate goal is $n_1 - n_2$. We are given $T_2 = T_1 =$
$T = 20°\text{C} + 273 = 293 \text{ K}$ and $V_2 = V_1 = V = \pi r^2 C = \pi(0.011 \text{ m}^2)(2.0 \text{ m}) = 7.60\times10^{-4} \text{ m}^3$.

Solve: We convert the gauge pressures to absolute pressures with $p = p_g + 1 \text{ atm} = p_g + 14.7 \text{ psi}$. $p_1 = 110 \text{ psi} +$
$14.7 \text{ psi} = 124.7 \text{ psi} = 860 \text{ kPa}$. $p_2 = 80 \text{ psi} + 14.7 \text{ psi} = 94.7 \text{ psi} = 653 \text{ kPa}$.

$$n_1 - n_2 = \frac{p_1 V_1}{RT_1} - \frac{p_2 V_2}{RT_2}$$

$$= \frac{V}{RT}(p_1 - p_2)$$

$$= \frac{7.60\times10^{-4} \text{ m}^3}{(8.31 \text{ J/mol}\cdot\text{K})(293 \text{ K})}(860 \text{ kPa} - 653 \text{ kPa})$$

$$= 0.0646 \text{ mol}$$

Thus 0.0646 mol of N_2 was lost; this is

$$0.0646 \text{ mol}\left(\frac{28 \text{ g}}{1 \text{ mol}}\right) = 1.8 \text{ g}$$

Assess: The result seems to be a reasonable number.

16.45. Model: Assume the trapped air to be an ideal gas.
Visualize:

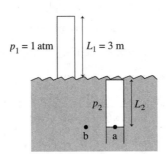

Initially, as the pipe is touched to the water surface and the gas inside is thus closed off from the air, the pressure $p_1 = p_{atmos} = 1$ atm and the volume is $V_1 = L_1 A$, where A is the cross-sectional area of the pipe. By pushing the pipe in *slowly*, the gas temperature in the pipe remains the same as the water temperature. Thus, this is an isothermal compression of the gas with $T_2 = T_1$.

Solve: From the ideal-gas law,

$$p_2 V_2 = p_1 V_1 \Rightarrow p_2 L_2 A = p_1 L_1 A \Rightarrow p_2 L_2 = p_{atmos} L_1 \Rightarrow p_2 = p_{atmos}(L_1/L_2)$$

As the pipe is pushed down, the increasing water pressure pushes water up into the pipe, compressing the air. In equilibrium, the pressure at points a and b, along a horizontal line, must be equal. (This is like the barometer. If the pressures at a and b weren't equal, the pressure difference would cause the liquid level in the pipe to move up or down.) The pressure at point a is just the gas pressure inside the pipe: $p_a = p_2$. The pressure at point b is the pressure at depth L_2 in water: $p_b = p_{atmos} + \rho g L_2$. Equating these gives

$$p_2 = p_{atmos} = \rho g L_2$$

Substituting the expression for p_2 from the ideal-gas equation above, the pressure equation becomes

$$\frac{p_{atmos} L_1}{L_2} = p_{atmos} + \rho g L_2 \Rightarrow \rho g L_2^2 + p_{atmos} L_2 - p_{atmos} L_1 = 0$$

This is a quadratic equation for L_2 with solutions

$$L_2 = \frac{-p_{atmos} \pm \sqrt{(p_{atmos})^2 + 4\rho g p_{atmos} L_1}}{2\rho g}$$

Length has to be a positive quantity, so the one physically acceptable solution is

$$L_2 = \frac{-101,300 \text{ Pa} + \sqrt{(101,300 \text{ Pa})^2 + 4(1000 \text{ kg/m}^3)(9.8 \text{ m/s}^2)(101,300 \text{ Pa})(3.0 \text{ m})}}{2(1000 \text{ kg/m}^3)(9.8 \text{ m/s}^2)} = 2.4 \text{ m}$$

16.47. Model: We assume that the volume of the tire and that of the air in the tire is constant.

Solve: A gauge pressure of 30 psi corresponds to an absolute pressure of $(30 \text{ psi}) + (14.7 \text{ psi}) = 44.7$ psi. Using the before-and-after relationship of an ideal gas for an isochoric (constant volume) process,

$$\frac{p_1}{T_1} = \frac{p_2}{T_2} \Rightarrow p_2 = \frac{T_2}{T_1} p_1 = \left(\frac{273 + 45}{273 + 15}\right)(44.7 \text{ psi}) = 49.4 \text{ psi}$$

Your tire gauge will read a gauge pressure $p_g = 49.4$ psi $- 14.7$ psi $= 34.7$ psi. ≈ 35 psi.

16.51. Model: The air in the closed section of the U-tube is an ideal gas.

Visualize: The length of the tube is $l = 1.0$ m and its cross-sectional area is A.

Solve: Initially, the pressure of the air in the tube is $p_1 = p_{atmos}$ and its volume is $V_1 = Al$. After the mercury is poured in, compressing the air, the air-pressure force supports the weight of the mercury. Thus the compressed pressure equals the pressure at the bottom of the column: $p_2 = p_{atmos} + \rho g L$. The volume of the compressed air is $V_2 = A(l - L)$. Because the mercury is poured in slowly, we will assume that the gas remains in thermal equilibrium with the surrounding air, so $T_2 = T_1$. In an isothermal process, pressure and volume are related by

$$p_1 V_1 = p_{atmos} Al = p_2 V_2 = (p_{atmos} + \rho g L) A(l - L)$$

Canceling the A, multiplying through, and solving for L gives

$$L = l - \frac{p_{atmos}}{\rho g} = 1.00 \text{ m} - \frac{101,300 \text{ Pa}}{(13,600 \text{ kg/m}^3)(9.8 \text{ m/s}^2)} = 0.24 \text{ m} = 24 \text{ cm}$$

16.53. Model: Assume that the compressed air in the cylinder is an ideal gas. The volume of the air in the cylinder is a constant.
Solve: Using the before-and-after relationship of an ideal gas,

$$\frac{p_2 V_2}{T_2} = \frac{p_1 V_1}{T_1} \Rightarrow p_2 = p_1 \frac{T_2}{T_1} \frac{V_1}{V_2} = (25 \text{ atm}) \left(\frac{1223 \text{ K}}{293 \text{ K}} \right) \frac{V_1}{V_1} = 104 \text{ atm}$$

where we have converted to the Kelvin temperature scale. Because the pressure does not exceed 110 atm, the compressed air cylinder does not blow.

16.55. Model: Assume that the gas is an ideal gas.
Solve:

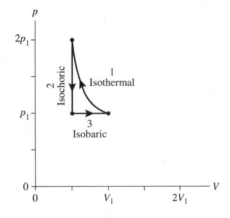

Assess: For the isothermal process, the pressure must double as the volume is halved. This is because p is proportional to $1/V$ for isothermal processes.

16.57. Model: Assume the nitrogen gas is an ideal gas.
Solve: (a) The number of moles of nitrogen is

$$n = \frac{M}{M_{mol}} = \frac{1 \text{ g}}{28 \text{ g/mol}} = \left(\frac{1}{28} \right) \text{mol}$$

Using the ideal-gas equation,

$$p_1 = \frac{nRT_1}{V_1} = \frac{(1/28 \text{ mol})(8.31 \text{ J/mol K})(298 \text{ K})}{(100 \times 10^{-6} \text{ m}^3)} = 8.84 \times 10^5 \text{ Pa} = 884 \text{ kPa} \approx 880 \text{ kPa}$$

(b) For the process from state 1 to state 3:

$$\frac{p_1 V_1}{T_1} = \frac{p_3 V_3}{T_3} \Rightarrow T_3 = T_1 \frac{p_3}{p_1} \frac{V_3}{V_1} = (298 \text{ K}) \left(\frac{1.5 p_1}{p_1} \right) \left(\frac{50 \text{ cm}^3}{100 \text{ cm}^3} \right) = 223.5 \text{ K} \approx -49°C$$

For the process from state 3 to state 2:

$$\frac{p_2 V_2}{T_2} = \frac{p_3 V_3}{T_3} \Rightarrow T_2 = T_3 \left(\frac{p_2}{p_3} \right) \left(\frac{V_2}{V_3} \right) = (223.5 \text{ K}) \left(\frac{2.0 p_1}{1.5 p_1} \right) \left(\frac{100 \text{ cm}^3}{50 \text{ cm}^3} \right) = 596 \text{ K} = 323°C$$

For the process from state 1 to state 4:

$$\frac{p_4 V_4}{T_4} = \frac{p_1 V_1}{T_1} \Rightarrow T_4 = T_1 \frac{p_4}{p_1} \frac{V_4}{V_1} = (298 \text{ K}) \left(\frac{1.5 p_1}{p_1} \right) \left(\frac{150 \text{ cm}^3}{100 \text{ cm}^3} \right) = 670.5 \text{ K} \approx 398°C$$

16.61. Model: We assume the oxygen gas is ideal.
Visualize: From the figure we glean $p_2 = 3p_1$ and $V_2 = 3V_1$. We are given $T_1 = 20°C + 273 = 293$ K.

Solve: Use the before-and-after version of the ideal-gas law.

$$\frac{p_1 V_1}{T_1} = \frac{p_2 V_2}{T_2}$$

$$T_2 = \frac{p_2 V_2}{p_1 V_1} T_1 = \frac{(3p_1)(3V_1)}{p_1 V_1} T_1 = 9T_1 = 9(293 \text{ K}) = 2637 \text{ K} = 2364°C$$

Assess: This is a hot temperature—higher than the melting point of many elements.

16.63. Model: The gas in the container is assumed to be an ideal gas.
Solve: (a) The gas starts at pressure $p_1 = 2.0$ atm, temperature $T_1 = 127°C = (127 + 273)$ K $= 400$ K and volume V_1. It is first compressed at a constant temperature $T_2 = T_1$ until $V_2 = \frac{1}{2} V_1$ and the pressure is p_2. It is then further compressed at constant pressure $p_3 = p_2$ until $V_3 = \frac{1}{2} V_2$. From the ideal-gas law,

$$\frac{p_2 V_2}{T_2} = \frac{p_1 V_1}{T_1} \Rightarrow p_2 = p_1 \frac{V_1}{V_2} \frac{T_2}{T_1} = (2.0 \text{ atm}) \frac{V_1}{\frac{1}{2} V_1} \times 1 = 4.0 \text{ atm}$$

Note that $T_2 = T_1 = 400$ K. Using the ideal-gas law once again,

$$\frac{p_3 V_3}{T_3} = \frac{p_2 V_2}{T_2} \Rightarrow T_3 = T_2 \frac{V_3}{V_2} \frac{p_3}{p_2} = (400 \text{ K}) \frac{\frac{1}{2} V_2}{V_2} \times 1 = 200 \text{ K} = -73°C$$

The final pressure and temperature are 4.0 atm and $-73°C$.
(b)

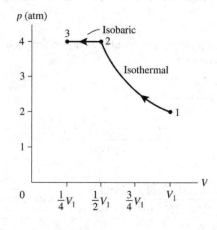

WORK, HEAT, AND THE FIRST LAW OF THERMODYNAMICS

Exercises and Problems

Section 17.1 It's All About Energy

Section 17.2 Work in Ideal-Gas Processes

17.1. Model: Assume the gas is ideal. The work done on a gas is the negative of the area under the pV curve.
Visualize: The gas is undergoing compression, so we expect the work done on it to be positive.
Solve: The work done on the gas is the negative of the area under the pV curve from V_i to V_f, which is the area of the rectangle and triangle. We have

$$W = -\int_{V_i}^{V_f} pdV = -\left[-(200\times10^{-6}\ m^3)(200\times10^3\ Pa) - \tfrac{1}{2}(200\times10^{-6}\ m^3)(200\times10^3\ Pa)\right] = 60\ J$$

where the inner negative signs come from the fact that $V_f < V_i$. The work done on the gas by the environment is thus $W = 60\ J$.
Assess: The environment does positive work on the gas to compress it.

Section 17.3 Heat

Section 17.4 The First Law of Thermodynamics

17.5. Visualize:

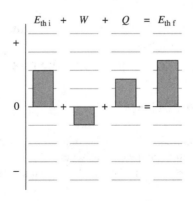

Solve: Because this is an isobaric process $W = -\int p\,dV = -p(V_f - V_i)$. Since V_f is greater than V_i, W is negative. That is, the gas expands. Since the final point is on a higher isotherm than the initial point, $T_f > T_i$. In other words, the thermal energy increases. For this to happen, the heat energy transferred into the gas must be larger than the work done by the gas.

17.9. Solve: This is an isobaric process. $W > 0$ because the gas is compressed, which transfers energy into the system. Also, 100 J of heat energy is transferred out of the gas. The first law of thermodynamics gives

$$\Delta E_{th} = W + Q = -p\Delta V + Q = -(4.0\times10^5 \text{ Pa})(200-600)\times10^{-6} \text{ m}^3 - 100 \text{ J} = 60 \text{ J}$$

Thermal energy increases by 60 J.

Section 17.5 Thermal Properties of Matter

17.11. Model: The addition of heat to the aluminum increases its thermal energy and its temperature.
Solve: The heat needed to change an object's temperature is $Q = Mc\Delta T$. The mass of the aluminum cube is

$$M = \rho_{Al}V = (2700 \text{ kg/m}^3)(0.060\times0.060\times0.060)\text{m}^3 = 0.5832 \text{ kg}$$

The specific heat of ice from Table 17.2 is $c_{Al} = 900$ J/kg K, so

$$Q = (0.5832 \text{ kg})(900 \text{ J/kg K})(100 \text{ K}) = 52 \text{ kJ}$$

Thus, the energy required to heat the Al block is 52 kJ.
Assess: The positive sign of Q means the system (i.e., Al block) gains energy.

17.13. Model: Heating the mercury changes its thermal energy and its temperature.
Solve: (a) The heat needed to change the mercury's temperature is

$$Q = Mc_{Hg}\Delta T \quad \Rightarrow \quad \Delta T = \frac{Q}{Mc_{Hg}} = \frac{100 \text{ J}}{(0.020 \text{ kg})(140 \text{ J/kg K})} = 35.7 \text{ K} \approx 36°C$$

(b) The amount of heat required to raise the temperature of the same amount of water by the same amount is
$$Q = Mc_{water}\Delta T = (0.020 \text{ kg})(4190 \text{ J/kg K})(35.7 \text{ K}) = 3000 \text{ J}$$

Assess: Q is directly proportional to c_{water} and the specific heat for water is much higher than the specific heat for mercury. This explains why $Q_{water} > Q_{mercury}$.

Section 17.6 Calorimetry

17.17. Model: We have a thermal interaction between the aluminum pan and the water.
Solve: The conservation of energy equation $Q_{Al} + Q_{water} = 0$ J gives

$$M_{Al}c_{Al}[T_f - (T_i)_{Al}] + M_{water}c_{water}[T_f - (T_i)_{water}]$$

The pan and water reach a common final temperature $T_f = 24.0°C$

$$(0.750 \text{ kg})(900 \text{ J/kg K})[24.0°C - (T_i)_{Al}] + (10.0\times10^{-3} \text{ m}^3)(1000 \text{ kg/m}^3)(4190 \text{ J/kg K})(24.0°C - 20.0°C)$$
$$= (675.0 \text{ J/K})[24.0°C - (T_i)_{Al}] + 167{,}600 \text{ J} = 0 \text{ J}$$
$$(T_i)_{Al} = 272°C = [(272)(9/5) + 32]°F = 522°F$$

Section 17.7 The Specific Heat of Gases

17.21. Model: Use the models of isochoric and isobaric heating. Note that the change in temperature on the Kelvin scale is the same as the change in temperature on the Celsius scale.
Solve: (a) The atomic mass number of argon is 40. That is, $M_{mol} = 40$ g/mol. The number of moles of argon gas in the container is

$$n = \frac{M}{M_{mol}} = \frac{1.0 \text{ g}}{40 \text{ g/mol}} = 0.025 \text{ mol}$$

The amount of heat is

$$Q = nC_V\Delta T = (0.025 \text{ mol})(12.5 \text{ J/mol K})(100°C) = 31.25 \text{ J} \approx 31 \text{ J}$$

(b) For the isobaric process $Q = nC_P\Delta T$ becomes

$$31.25 \text{ J} = (0.025 \text{ mol})(20.8 \text{ J/mol K})\Delta T \quad \Rightarrow \quad \Delta T = 60°C$$

17.23. Model: N_2 is a diatomic ideal gas. It condenses at $-196°C$ (see Table 17.3).
Solve: The number of moles of nitrogen is

$$n = \frac{M}{M_{mol}} = \frac{7.0 \text{ g}}{28 \text{ g/mol}} = 0.25 \text{ mol}$$

Less energy is required to cool the gas in an isochoric process because no work is done, so the minimum heat that must be lost by the gas to cool it to the condensation point is

$$Q = nC_V\Delta T = (0.25 \text{ mol})(20.8 \text{ J/mol K})(-196°C - 20°C) = -1123 \text{ J}$$

To condense this to a liquid, the following additional heat of vaporization must be removed:

$$Q = -ML_v = -(0.0070 \text{ kg})(1.99 \times 10^5 \text{ J/kg}) = -1393 \text{ J}$$

Thus, the total heat that must be *removed* from the gas is $1123 \text{ J} + 1393 \text{ J} = 2.5 \text{ kJ}$, where we have taken positive sign because this is the heat that we (the environment) must remove.
Assess: Because the gas loses this heat, the sign of its heat transfer is negative.

17.25. Model: We assume the gas is an ideal gas and $\gamma = 1.40$ for a diatomic gas.
Solve: Using the ideal-gas law,

$$V_i = \frac{nRT_i}{p_i} = \frac{(0.10 \text{ mol})(8.31 \text{ J/mol K})(423 \text{ K})}{(3 \times 1.013 \times 10^5 \text{ Pa})} = 1.157 \times 10^{-3} \text{ m}^3$$

(a) For an adiabatic process,

$$p_i V_i^\gamma = p_f V_f^\gamma$$

$$V_f = V_i \left(\frac{p_i}{p_f}\right)^{1/\gamma} = (1.157 \times 10^{-3} \text{ m}^3)\left(\frac{p_i}{0.5 \, p_i}\right)^{1/1.40} = 1.9 \times 10^{-3} \text{ m}^3$$

(b) To find the final temperature, we use the ideal-gas law once again as follows:

$$T_f = T_i \frac{p_f \, V_f}{p_i \, V_i} = (423 \text{ K})\left(\frac{0.5 \, p_i}{p_i}\right)\left(\frac{1.90 \times 10^{-3} \text{ m}^3}{1.157 \times 10^{-3} \text{ m}^3}\right) = 346.9 \text{ K} \approx 74°C$$

Section 17.8 Heat-Transfer Mechanisms

17.29. Model: Assume the lead sphere is an ideal radiator with $e = 1$. Also assume that the highest temperature the solid lead sphere can have is the melting temperature of lead.
Visualize: Use Equation 17.47. First look up the melting temperature of lead in Table 17.3: $T_m = 328°C = 601 \text{ K}$.
Then compute the surface area of the sphere: $A = 4\pi R^2 = 4\pi(0.050 \text{ m})^2 = 0.0314 \text{ m}^2$
Solve:

$$\frac{Q}{\Delta t} = e\sigma A T^4 = (1)(5.67 \times 10^{-8} \text{ W/m}^2 \text{ K}^4)(0.0314 \text{ m}^2)(601 \text{ K})^4 = 230 \text{ W}$$

Assess: If the sphere were larger it could radiate more power without melting.

17.33. Model: The potential energy of the bowling ball is transferred into thermal energy of the mixture. We assume the starting temperature of the bowling ball to be $0°C$.
Solve: The potential energy of the bowling ball is

$$U_g = M_{ball}gh = (11 \text{ kg})(9.8 \text{ m/s}^2)h = (107.8 \text{ kg m/s}^2)h$$

This energy is transferred into the mixture of ice and water and melts 5 g of ice. That is,

$$(107.8 \text{ kg m/s}^2)h = \Delta E_{th} = M_w L_f \implies h = \frac{(0.005 \text{ kg})(3.33 \times 10^5 \text{ J/kg})}{(107.8 \text{ kg m/s}^2)} = 15 \text{ m}$$

17.35. Model: Model the alligator as a rectangular solid with an upper surface $2.9 \times 0.60 \text{ m}^2 = 1.74 \text{ m}^2$.
Solve: Apply Equation 17.18 to find the heat required to warm the alligator by $7°C$:

$$Q = Mc\Delta T = (350 \text{ kg})(3400 \text{ J/kg K})(7 \text{ K}) = 8.33 \text{ MJ}$$

To acquire this energy from sunlight would take approximately

$$t = \frac{Q}{P} = \frac{8.33 \text{ MJ}}{(500 \text{ W/m}^2)(1.74 \text{ m}^2)} = (9.57 \times 10^4 \text{ s})\left(\frac{1 \text{ min}}{60 \text{ s}}\right)\left(\frac{1 \text{ h}}{60 \text{ min}}\right) = 2.7 \text{ h}$$

17.37. Model: Model evaporation in the same manner as for the previous problem.
Solve: At 12 breaths/min, we exhale every minute $M = (12)(25 \times 10^{-6} \text{ kg}) = 300 \times 10^{-6} \text{ kg}$ of water. To vaporize $300 \ \mu g$ of water at $35°C$ every minute requires an approximate power of

$$P_2 = \frac{Q_2}{(60 \text{ s})} = \frac{ML_v}{(60 \text{ s})} = \frac{(300 \times 10^{-6} \text{ kg})(24 \times 10^5 \text{ J/kg})}{60 \text{ s}} = 12 \text{ J/s}$$

Assess: This is much less than the energy lost from the body by radiation.

17.39. Model: There are three interacting systems: aluminum, copper, and ethyl alcohol.
Solve: The aluminum, copper, and alcohol form a closed system, so $Q = Q_{Al} + Q_{cu} + Q_{eth} = 0 \text{ J}$. The mass of the alcohol is

$$M_{eth} = \rho V = (790 \text{ kg/m}^3)(50 \times 10^{-6} \text{ m}^3) = 0.0395 \text{ kg}$$

Expressed in terms of specific heats and using the fact that $\Delta T = T_f - T_i$, the $Q = 0 \text{ J}$ condition is

$$M_{Al}c_{Al}\Delta T_{Al} + M_{Cu}c_{Cu}\Delta T_{Cu} + M_{eth}c_{eth}\Delta T_{eth} = 0 \text{ J}$$

Substituting the appropriate values into this expression gives

$$(0.010 \text{ kg})(900 \text{ J/kg K})(298 \text{ K} - 473 \text{ K}) + (0.020 \text{ kg})(385 \text{ J/kg K})(298 \text{ K} - T)$$
$$+ (0.0395 \text{ kg})(2400 \text{ J/kg K})(298 \text{ K} - 288 \text{ K}) = -1575 \text{ J} + (7.7 \text{ J/K})(298 - T) + 948 \text{ J} = 0 \text{ J}$$
$$T = 216.6 \text{ K} = -56.4°C \approx -56°C$$

17.43. Solve: From Table 17.4, we find the specific heat of O_2 at constant volume is 20.9 J/mol K. The atomic number of oxygen is 16, so O_2 has 32 g/mol or 0.032 kg/mol. Converting the units gives the specific heat in J/kg K:

$$c_V = \frac{20.9 \text{ J/mol K}}{0.032 \text{ kg/mol}} = 650 \text{ J/kg K}$$

17.47. Model: There are two interacting systems: the nuclear reactor and the water. The heat generated by the nuclear reactor is used to raise the water temperature.
Solve: For the closed reactor-water system, energy conservation per second requires

$$Q = Q_{reactor} + Q_{water} = 0 \text{ J}$$

The heat from the reactor in $\Delta t = 1 \text{ s}$ is $Q_{reactor} = -2000 \text{ MJ} = -2.0 \times 10^9 \text{ J}$ and the heat absorbed by the water is

$$Q_{water} = m_{water}c_{water}\Delta T = m_{water}(4190 \text{ J/kg K})(12 \text{ K})$$

$$-2.0\times10^9 \text{ J} + m_{water}(4190 \text{ J/kg K})(12 \text{ K}) = 0 \text{ J} \implies m_{water} = 3.98\times10^4 \text{ kg}$$

Each second, 3.98×10^4 kg of water is needed to remove heat from the nuclear reactor. Thus, the water flow per minute is

$$\left(3.98\times10^4 \frac{\text{kg}}{\text{s}}\right)\left(\frac{60 \text{ s}}{\text{min}}\right)\left(\frac{1 \text{ m}^3}{1000 \text{ kg}}\right)\left(\frac{1 \text{ L}}{10^{-3} \text{ m}^3}\right) = 2.4\times10^6 \text{ L/min}$$

Thus, 2.4×10^6 L of cooling water must flow through the reactor each minute.

17.49. Model: These are isothermal and isobaric ideal-gas processes.
Solve: (a) The work done at constant temperature is

$$W = -\int_{V_i}^{V_f} p\,dV = -\int_{V_i}^{V_f} \frac{nRT}{V}\,dV = -nRT(\ln V_f - \ln V_i) = -nRT\ln(V_f/V_i)$$

$$= -(2.0 \text{ mol})(8.31 \text{ J/mol K})(303 \text{ K})\ln\left(\tfrac{1}{3}\right) = 5.5 \text{ kJ}$$

(b) The work done at constant pressure is

$$W = -\int_{V_i}^{V_f} p\,dV = -p(V_f - V_i) = -p\left(\frac{V_i}{3} - V_i\right) = \frac{2}{3}pV_i$$

$$= \frac{2}{3}nRT = \frac{2}{3}(2.0 \text{ mol})(8.31 \text{ J/mol K})(303 \text{ K}) = 3.4 \text{ kJ}$$

(c) For an isothermal process in which $V_f = \frac{1}{3}V_i$, the pressure changes to $p_f = 3p_i = 4.5$ atm.

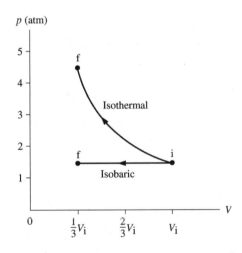

17.53. Model: The process in part (a) is isochoric and the process in part (b) is isobaric.
Solve: (a) Initially $V_1 = (0.20 \text{ m})^3 = 0.0080 \text{ m}^3 = 8.0 \text{ L}$ and $T_1 = 293 \text{ K}$. Helium has an atomic mass number $A = 4$, so 3 g of helium is $n = M/M_{mol} = 0.75$ mole of helium. We can find the initial pressure from the ideal-gas law:

$$p_1 = \frac{nRT_1}{V_1} = \frac{(0.75 \text{ mol})(8.31 \text{ J/mol K})(293 \text{ K})}{0.0080 \text{ m}^3} = 228 \text{ kPa} = 2.25 \text{ atm}$$

Heating the gas will raise its temperature. A constant-volume process has $Q = nC_V\Delta T$, so

$$\Delta T = \frac{Q}{nC_V} = \frac{1000 \text{ J}}{(0.75 \text{ mol})(12.5 \text{ J/mol K})} = 107 \text{ K}$$

This raises the final temperature to $T_2 = T_1 + \Delta T = 400 \text{ K}$. Because the process is isochoric,

$$\frac{p_2}{T_2} = \frac{p_1}{T_1} \quad \Rightarrow \quad p_2 = \frac{T_2}{T_1} p_1 = \frac{400 \text{ K}}{293 \text{ K}} (2.25 \text{ atm}) = 3.1 \text{ atm}$$

(b) The initial conditions are the same as part a, but now $Q = nC_p \Delta T$. Thus,

$$\Delta T = \frac{Q}{nC_p} = \frac{1000 \text{ J}}{(0.75 \text{ mol})(20.8 \text{ J/mol K})} = 64.1 \text{ K}$$

Now the final temperature is $T_2 = T_1 + \Delta T = 357 \text{ K}$. Because the process is isobaric,

$$\frac{V_2}{T_2} = \frac{V_1}{T_1} \quad \Rightarrow \quad V_2 = \frac{T_2}{T_1} V_1 = \frac{357 \text{ K}}{293 \text{ K}} (0.0080 \text{ m}^3) = 0.0097 \text{ m}^3 = 9.7 \text{ L}$$

(c)

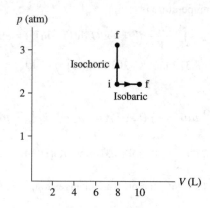

17.55. Model: This is an isothermal process.
Solve: (a) The final temperature is $T_2 = T_1$ because the process is isothermal.
(b) The work done on the gas is

$$W = -\int_{V_1}^{V_2} p\,dV = -\int_{V_1}^{V_2} \frac{nRT_1}{V}\,dV = -nRT_1 \ln\frac{V_2}{V_1} = -nRT_1 \ln 2$$

(c) From the first law of thermodynamics $\Delta E_{th} = W + Q = 0$ J because $\Delta T = 0$ K. Thus, the heat energy transferred to the gas is $Q = -W = nRT_1 \ln 2$.

17.57. Model: The gas is an ideal gas.
Visualize: In the figure, call the upper-left corner (on process A) 2 and the lower-right corner (on process B) 3.
Solve: The change in thermal energy is the same for any gas process that has the same ΔT. Processes A and B have the same ΔT, since they start and end at the same points, so $(\Delta E_{th})_A = (\Delta E_{th})_B$. The first law then gives

$$(\Delta E_{th})_A = Q_A + W_A = (\Delta E_{th})_B = Q_B + W_B \quad \Rightarrow \quad Q_A - Q_B = W_B - W_A$$

In process B, work $W = -p\Delta V = -p_i(2V_i - V_i) = -p_iV_i$ is done during the isobaric process $i \rightarrow 3$. No work is done during the isochoric process $3 \rightarrow f$. Thus $W_B = -p_iV_i$. Similarly, no work is done during the isochoric process $i \rightarrow 2$ of process A, but $W = -p\Delta V = -2p_i(2V_i - V_i) = -2p_iV_i$ is done during the isobaric process $2 \rightarrow f$. Thus $W_A = -2p_iV_i$. Combining these gives

$$Q_A - Q_B = W_B - W_A = -p_iV_i - (-2p_iV_i) = p_iV_i$$

17.61. Model: Model the gas as an ideal gas and the compression as an adiabatic process.
Solve: The gas temperature changes as the cylinder volume changes. Apply Equation 17.39:

$$TV^{\gamma-1} = T_iV_i^{\gamma-1} \quad \Rightarrow \quad \ln\left(\frac{T}{T_i}\right) = (\gamma-1)\ln\left(\frac{V_i}{V}\right) = (\gamma-1)\ln\left(\frac{L_i}{L}\right)$$

The cross section area of the cylinder doesn't change, so we could replace the ratio of the volumes with the ratio of the cylinder lengths. The initial values are $L_i = 20$ cm and $T_i = 21°C = 294$ K (room temperature). Note that the data show how far the piston is pushed; the length is $L = 20$ cm minus the push distance. If we plot $\ln(T/T_i)$ versus $\ln(L_i/L)$, the graph should be a straight line with slope $\gamma - 1$. The plot is shown below. The experimental slope of 0.29 gives $\gamma = 1.29$.

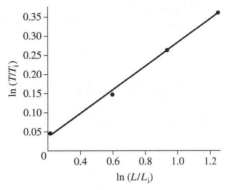

17.63. Model: The gas is assumed to be an ideal gas that is subjected to isobaric and isochoric processes.
Solve: (a) The initial conditions are $p_1 = 3.0$ atm $= 304,000$ Pa, $V_1 = 100$ cm$^3 = 1.0 \times 10^{-4}$ m^3, and $T_1 = 100°C = 373$ K. The number of moles of gas is

$$n = \frac{p_1 V_1}{R T_1} = \frac{(304,000 \text{ Pa})(1.0 \times 10^{-4} \text{ m}^3)}{(8.31 \text{ J/mol K})(373 \text{ K})} = 9.81 \times 10^{-3} \text{ mol}$$

At point 2 we have $p_2 = p_1 = 3.0$ atm and $V_2 = 300$ cm$^3 = 3V_1$. This is an isobaric process, so

$$\frac{V_2}{T_2} = \frac{V_1}{T_1} \quad \Rightarrow \quad T_2 = \frac{V_2}{V_1} T_1 = 3(373 \text{ K}) = 1119 \text{ K}$$

The gas is heated to raise the temperature from T_1 to T_2. The amount of heat required is

$$Q = n C_P \Delta T = (9.81 \times 10^{-3} \text{ mol})(20.8 \text{ J/mol K})(1119 \text{ K} - 373 \text{ K}) = 0.15 \text{ kJ}$$

This amount of heat is *added* during process $1 \to 2$.
(b) Point 3 returns to $T_3 = 100°C = 373$ K. This is an isochoric process, so

$$Q = n C_V \Delta T = (9.81 \times 10^{-3} \text{ mol})(12.5 \text{ J/mol K})(373 \text{ K} - 1119 \text{ K}) = -91 \text{ J}$$

This amount of heat is *removed* during process $2 \to 3$.

17.67. Model: Model the gas as an ideal gas.
Solve: Because the two gases have the same initial conditions, the ideal-gas law tells us that they have the same number of moles. Apply Equation 17.25 for constant pressure to find the heat energy required by the first process: $Q = n C_P \Delta T_1$. Insert this into Equation 17.25 for a constant volume process to find the temperature increase:

$$Q = n C_V \Delta T_2 \quad \Rightarrow \quad \Delta T_2 = \frac{Q}{n C_V} = \frac{n C_P \Delta T_1}{n C_V} = \frac{C_P}{C_V} \Delta T_1 = \gamma \Delta T_1$$

For a diatomic gas, $\lambda = 1.40$ (see Equation 17.35), so the temperature change for the second container of gas is

$$\Delta T_2 = 1.40(20°C) = 28°C$$

17.69. Model: The gas is assumed to be an ideal gas that is subjected to an isochoric process.
Solve: (a) The number of moles in 14.0 g of N_2 gas is

$$n = \frac{M}{M_{mol}} = \frac{14.0 \text{ g}}{28 \text{ g/mol}} = 0.50 \text{ mol}$$

At $T_i = 273$ K and $p_i = 1.0$ atm $= 1.013 \times 10^5$ Pa, the gas has a volume

$$V_i = \frac{nRT_i}{p_i} = 0.0112 \text{ m}^3 = 11.2 \text{ L}$$

For an isochoric process $(V_i = V_f)$,

$$\frac{T_f}{T_i} = \frac{p_f}{p_i} = \frac{20 \text{ atm}}{1 \text{ atm}} = 20 \quad \Rightarrow \quad T_f = 20(273 \text{ K}) = 5.5 \text{ kK}$$

(b) The work done on the gas is $W = -p\Delta V = 0$ J.

(c) The heat input to the gas is

$$Q = nC_V(T_f - T_i) = (0.50 \text{ mol})(20.8 \text{ J/mol K})(5460 \text{ K} - 273 \text{ K}) = 5.4 \times 10^4 \text{ J}$$

(d) The pressure ratio is

$$\frac{p_{max}}{p_{min}} = \frac{p_f}{p_i} = \frac{20 \text{ atm}}{1 \text{ atm}} = 20$$

(e)

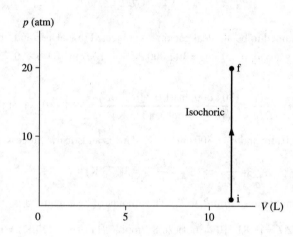

THE MICRO/MACRO CONNECTION

Exercises and Problems

Section 18.1 Molecular Speeds and Collisions

18.1. Solve: The volume of the nitrogen gas is 1.0 m^3 and its temperature is 20°C or 293 K. The number of gas molecules can be found as

$$N = nN_A = \frac{pV}{RT}N_A = \frac{(1.013 \times 10^5 \text{ Pa})(1.0 \text{ m}^3)}{(8.31 \text{ J/mol K})(293 \text{ K})}(6.02 \times 10^{23} \text{ mol}^{-1}) = 2.5 \times 10^{25}$$

According to Figure 18.2, 12% of the molecules have a speed between 700 and 800 m/s, 7% between 800 and 900 m/s, and 3% between 900 and 1000 m/s. Thus, the number of molecules in the cube with a speed between 700 m/s and 1000 m/s is $(0.22)(2.51 \times 10^{25}) = 5.5 \times 10^{24}$.

18.5. Solve: (a) The mean free path of a molecule in a gas at temperature T_1, volume V_1, and pressure p_1 is $\lambda_1 = 300 \text{ nm}$. We also know that

$$\lambda = \frac{1}{4\sqrt{2}\pi(N/V)r^2} \Rightarrow \lambda \propto V$$

Although $T_2 = 2T_1$, constant volume $(V_2 = V_1)$ means that $\lambda_2 = \lambda_1 = 300 \text{ nm}$.

(b) For $T_2 = 2T_1$ and $p_2 = p_1$, the ideal gas equation gives

$$\frac{p_1V_1}{Nk_BT_1} = \frac{p_2V_2}{Nk_BT_2} = \frac{p_1V_2}{Nk_B(2T_1)} \Rightarrow V_2 = 2V_1$$

Because $\lambda \propto V$, $\lambda_2 = 2\lambda_1 = 2(300 \text{ nm}) = 600 \text{ nm}$.

18.7. Solve: The number density of the Ping-Pong balls inside the box is

$$\frac{N}{V} = \frac{2000}{1.0 \text{ m}^3} = 2000 \text{ m}^{-3}$$

With $r = (3.0 \text{ cm})/2 = 1.5 \text{ cm}$, the mean free path of the balls is

$$\lambda = \frac{1}{4\sqrt{2}\pi(N/V)(r^2)} = 0.125 \text{ m} = 12.5 \text{ cm} \approx 13 \text{ cm}$$

Section 18.2 Pressure in a Gas

18.9. Solve: (a) In tabular form we have

Particle	v_x (m/s)	v_y (m/s)	v_x^2 (m/s)^2	v_y^2 (m/s)^2	v^2 (m/s)^2	v(m/s)
1	20	−30	400	900	1300	36.06
2	40	70	1600	4900	6500	80.62
3	−80	20	6400	400	6800	82.46
4	30	0	900	0	900	30.00
5	40	−40	1600	1600	3200	56.57
6	−50	−20	2500	400	2900	53.85
Average	0	0			3600	56.59

The average velocity is $\vec{v}_{avg} = 0\,\hat{i} + 0\,\hat{j}$.

(b) The average speed is $v_{avg} = 57$ m/s.

(c) The root-mean-square speed is $v_{rms} = \sqrt{(v^2)_{avg}} = \sqrt{3600 \text{ m}^2/\text{s}^2} = 60$ m/s.

18.13. Model: Pressure is due to random collisions of gas molecules with the walls.
Solve: According to Equation 18.8, the collision rate with one wall is

$$\text{rate of collisions} = \frac{N_{coll}}{\Delta t_{coll}} = \frac{F_{net}}{2mv_x} = \frac{pA}{2mv_x}$$

where $F_{net} = pA$ is the force exerted on area A by the gas pressure. However, this equation assumed that all molecules are moving in the x-direction with constant speed. The rms speed v_{rms} is for motion in three dimensions at varying speeds. Consequently, we need to replace v_x not with $(v_x)_{avg}$, which is zero, but with

$$v_x \rightarrow \sqrt{(v_x^2)_{avg}} = \sqrt{\frac{v_{rms}^2}{3}} = \frac{v_{rms}}{\sqrt{3}}$$

With this change,

$$\text{rate of collisions} = \frac{\sqrt{3}pA}{2mv_{rms}} = \frac{\sqrt{3}(2\times101{,}300 \text{ Pa})(0.10 \text{ m}\times0.10 \text{ m})}{2(28\times1.661\times10^{-27} \text{ kg})(576 \text{ m/s})} = 6.5\times10^{25} \text{ s}^{-1}$$

This collision rate can also be found by using the expression in Equation 18.10, making the same change in v_x, and using the ideal-gas law to determine N/V.

Section 18.3 Temperature

18.17. Solve: The average translational kinetic energy per molecule is

$$\epsilon_{avg} = \frac{1}{2}mv_{rms}^2 = \frac{3}{2}k_BT \Rightarrow v_{rms} = \sqrt{\frac{3k_BT}{m}}$$

Since we want the v_{rms} for H_2 and N_2 to be equal,

$$\sqrt{\frac{3k_BT_{H_2}}{m_{H_2}}} = \sqrt{\frac{3k_BT}{m_{N_2}}} \Rightarrow T_{H_2} = \frac{m_{H_2}}{m_{N_2}}T_{N_2} = \left(\frac{2 \text{ u}}{28 \text{ u}}\right)(373 \text{ K}) = 27 \text{ K} = -246°C$$

18.19. Solve: Solve Equation 18.18 for v_{rms} to see that if both p and V are halved then v_{rms} is also halved.

$$v_{rms} = \sqrt{\frac{3pV}{Nm}}$$

So the new rms speed will be 300 m/s.
Assess: Think microscopically; for the pressure to double when the volume is doubled the particles will have to be going a lot faster and hit the walls more often.

18.23. Solve: (a) The average translational kinetic energy per molecule is

$$K_{avg} = \frac{1}{2}mv_{rms}^2 = \frac{3}{2}k_BT$$

This means K_{avg} doubles if the temperature T doubles.

(b) The root-mean-square speed v_{rms} increases by a factor of $\sqrt{2}$ as the temperature doubles.

(c) The mean free path is

$$\lambda = \frac{1}{4\sqrt{2}\pi(N/V)r^2}$$

Because N/V and r do not depend on T, doubling temperature has no effect on λ.

18.25. Solve: (a) The average kinetic energy of a proton at the center of the sun is

$$K_{avg} = \frac{3}{2}k_BT \approx \frac{3}{2}(1.38\times10^{-23} \text{ J/K})(2.0\times10^7 \text{ K}) = 4.1\times10^{-16} \text{ J}$$

(b) The root-mean-square speed of the proton is

$$v_{rms} = \sqrt{\frac{3k_BT}{m}} \approx \sqrt{\frac{3(1.38\times10^{-23} \text{ J/K})(2.0\times10^7 \text{ K})}{1.67\times10^{-27} \text{ kg}}} = 7.0\times10^5 \text{ m/s}$$

Section 18.4 Thermal Energy and Specific Heat

18.29. Solve: The volume of the air is $V = 6.0 \text{ m}\times8.0 \text{ m}\times3.0 \text{ m} = 144.0 \text{ m}^3$, the pressure $p = 1 \text{ atm} = 1.013\times10^5 \text{ Pa}$, and the temperature $T = 20°C = 293 \text{ K}$. The number of moles of the gas is

$$n = \frac{pV}{RT} = 5991 \text{ mol}$$

This means the number of molecules is

$$N = nN_A = (5991 \text{ mols})(6.022\times10^{23} \text{ mol}^{-1}) = 3.61\times10^{27} \text{ molecules}$$

Since air is a diatomic gas, the room's thermal energy is

$$E_{th} = N\epsilon_{avg} = N\left(\frac{5}{2}k_BT\right) = 3.6\times10^7 \text{ J}$$

Assess: The room's thermal energy can also be obtained as follows:

$$E_{th} = nC_vT = (5991 \text{ mol})(20.8 \text{ J/mol K})(293 \text{ K}) = 3.6\times10^7 \text{ J}$$

18.33. Solve: (a) Nitrogen is a diatomic gas. The thermal energy of the gas at $T = 20°C = 293 \text{ K}$ is

$$E_{th} = \frac{5}{2}nRT = \left(\frac{pV}{RT}\right)\left(\frac{5}{2}R\right)T$$

$$= \frac{(1.013\times10^7 \text{ Pa})(15,000\times10^{-6} \text{ m}^3)}{(8.31 \text{ J/mol K})(293 \text{ K})}(20.8 \text{ J/mol K})(293 \text{ K}) = 3.80\times10^5 \text{ J}$$

(b) The number density of the gas is

$$\frac{N}{V} = \frac{p}{k_BT} = \frac{(1.013\times10^7 \text{ Pa})}{(1.38\times10^{-23} \text{ J/K})(293 \text{ K})} = 2.51\times10^{27} \text{ m}^{-3}$$

Since nitrogen is a diatomic molecule, the mean free path is

$$\lambda = \frac{1}{4\sqrt{2}\pi(N/V)r^2} = \frac{1}{4\sqrt{2}\pi(2.51\times10^{27} \text{ m}^{-3})(1.0\times10^{-10} \text{ m})^2} = 2.25\times10^{-9} \text{ m}$$

(c) For an isothermal process, $\Delta T = 0 \text{ K}$. Thus $\Delta E_{th} = 0 \text{ J}$.

Section 18.5 Thermal Interactions and Heat

18.35. Visualize: Because gas A starts with 9000 J of energy and transfers 1000 J to gas B then $E_{Af} = 8000 J$. We are also given $n_A = 4.0$ mol and $n_B = 3.0$ mol.
Solve:

$$E_{Af} = \frac{n_A}{n_A + n_B} E_{tot} \Rightarrow E_{tot} = \frac{n_A + n_B}{n_A} E_{Af}$$

The final total energy is the same as the initial total energy, so

$$E_{Bi} = E_{tot} - E_{Ai} = \frac{n_A + n_B}{n_A} E_{Af} - E_{Ai} = \left(\frac{4.0 \text{ mol} + 3.0 \text{ mol}}{4.0 \text{ mol}} \right)(8000 \text{ J}) - 9000 \text{ J} = 5000 \text{ J}$$

Assess: We can also compute $E_{Bf} = 6000$ J showing E_{tot} is the same before and after.

18.39. Solve: (a) To identify the gas, we need to determine its atomic mass number A or, equivalently, the mass m of each atom or molecule. The mass density ρ and the number density (N/V) are related by $\rho = m(N/V)$, so the mass is $m = \rho(V/N)$. From the ideal-gas law, the number density is

$$\frac{N}{V} = \frac{p}{kT} = \frac{50{,}000 \text{ Pa}}{(1.38 \times 10^{-23} \text{ J/K})(300 \text{ K})} = 1.208 \times 10^{25} \text{ m}^{-3}$$

Thus, the mass of an atom is

$$m = \rho \frac{V}{N} = \frac{8.02 \times 10^{-2} \text{ kg/m}^3}{1.208 \times 10^{25} \text{ m}^{-3}} = 6.64 \times 10^{-27} \text{ kg}$$

Converting to atomic mass units,

$$A = 6.64 \times 10^{-27} \text{ kg} \times \frac{1 \text{ u}}{1.661 \times 10^{-27} \text{ kg}} = 4.00 \text{ u}$$

This is the atomic mass of helium.
(b) Knowing the mass, we find v_{rms} to be

$$v_{rms} = \sqrt{\frac{3k_B T}{m}} = \sqrt{\frac{3(1.38 \times 10^{-23} \text{ J/K})(300 \text{ K})}{6.64 \times 10^{-27} \text{ kg}}} = 1370 \text{ m/s}$$

(c) A typical atomic radius is $r \approx 0.5 \times 10^{-10}$ m. The mean free path is thus

$$\lambda = \frac{1}{4\sqrt{2}\pi(N/V)r^2} = \frac{1}{4\sqrt{2}\pi(1.208 \times 10^{25} \text{ m}^{-3})(0.5 \times 10^{-10} \text{ m})^2} = 1.86 \times 10^{-6} \text{ m} = 1.86 \text{ } \mu m$$

18.41. Solve: (a) The number density is $N/V = 1 \text{ cm}^{-3} = 10^6 \text{ m}^{-3}$. Using the ideal-gas equation,

$$p = \frac{N}{V} k_B T \approx (1 \times 10^6 \text{ m}^{-3})(1.38 \times 10^{-23} \text{ J/K})(3 \text{ K})$$

$$= 4 \times 10^{-17} \text{ Pa} \times \frac{1 \text{ atm}}{1.013 \times 10^5 \text{ Pa}} = 4 \times 10^{-22} \text{ atm}$$

(b) For a monatomic gas,

$$v_{rms} = \sqrt{\frac{3k_B T}{m}} = \sqrt{\frac{3(1.38 \times 10^{-23} \text{ J/K})(3 \text{ K})}{1.67 \times 10^{-27} \text{ kg}}} = 270 \text{ m/s}$$

(c) The thermal energy is $E_{th} = \frac{3}{2} N k_B T$, where $N = (10^6 \text{ m}^{-3})V$. Thus

$$E_{th} = 1.0 \text{ J} = \frac{3}{2}(10^6 \text{ m}^{-3})V(1.38 \times 10^{-23} \text{ J/K})(3 \text{ K}) \Rightarrow V = 1.6 \times 10^{16} \text{ m}^3 = L^3$$

$$\Rightarrow L = 2.5 \times 10^5 \text{ m}$$

18.43. Solve: Fluorine has atomic mass number $A = 19$. Thus the root-mean-square speed of $^{238}UF_6$ is

$$v_{rms}(^{238}UF_6) = \sqrt{\frac{3k_B T}{m}} = \sqrt{\frac{3k_B T}{238\,u + 6\times19\,u}}$$

The ratio of the root-mean-square speed for the molecules of this isotope and the $^{235}UF_6$ molecules is

$$\frac{v_{rms}(^{235}UF_6)}{v_{rms}(^{238}UF_6)} = \sqrt{\frac{(238+6\times19)\,u}{(235+6\times19)\,u}} = \sqrt{\frac{352}{349}} = 1.0043$$

18.45. Solve: The pressure on the wall with area $A = 10\text{ cm}^2 = 10\times10^{-4}\text{ m}^2$ is

$$p = \frac{F}{A} = \frac{\Delta(mv)N}{A\Delta t}$$

where $K_{esc} = \frac{1}{2}mv_{esc}^2$ is the number of N_2 molecules colliding with the wall every second and $\Delta(mv)$ is the change in momentum for one collision. The mass of the nitrogen molecule is

$$m = 28\,u = 28(1.66\times10^{-27}\text{ kg}) = 4.648\times10^{-26}\text{ kg}$$

and $\Delta v = 400\text{ m/s} - (-400\text{ m/s}) = 800\text{ m/s}$. Thus,

$$p = \frac{(4.648\times10^{-26}\text{ kg})(800\text{ m/s})(5.0\times10^{23}\text{ s}^{-1})}{1.0\times10^{-3}\text{ m}^2} = 1.9\times10^4\text{ Pa}$$

18.49. Solve: (a) The number of moles of helium and oxygen are

$$n_{helium} = \frac{2.0\text{ g}}{4.0\text{ g/mol}} = 0.50\text{ mol} \qquad n_{oxygen} = \frac{8.0\text{ g}}{32.0\text{ g/mol}} = 0.25\text{ mol}$$

Since helium is a monoatomic gas, the initial thermal energy is

$$E_{helium\,i} = n_{helium}\left(\tfrac{3}{2}RT_{helium}\right) = (0.50\text{ mol})\left(\tfrac{3}{2}\right)(8.31\text{ J/mol K})(300\text{ K}) = 1870\text{ J} \approx 1900\text{ J}$$

Since oxygen is a diatomic gas, the initial thermal energy is

$$E_{oxygen\,i} = n_{oxygen}\left(\tfrac{5}{2}RT_{oxygen}\right) = (0.25\text{ mol})\left(\tfrac{5}{2}\right)(8.31\text{ J/mol K})(600\text{ K}) = 3116\text{ J} \approx 3100\text{ J}$$

(b) The total initial thermal energy is

$$E_{tot} = E_{helium\,i} + E_{oxygen\,i} = 4986\text{ J}$$

As the gases interact, they come to equilibrium at a common temperature T_f. This means

$$4986\text{ J} = n_{helium}\left(\tfrac{3}{2}RT_f\right) + n_{oxygen}\left(\tfrac{5}{2}RT_f\right)$$

$$\Rightarrow T_f = \frac{4986\text{ J}}{\left(\tfrac{1}{2}R\right)(3n_{helium}+5n_{oxygen})} = \frac{4986\text{ J}}{\tfrac{1}{2}(8.31\text{ J/mol K})(3\times0.50\text{ mol}+5\times0.25\text{ mol})} = 436.4\text{ K} = 436\text{ K}$$

The thermal energies at the final temperature T_f are

$$E_{helium\,f} = n_{helium}\left(\tfrac{3}{2}RT_f\right) = \left(\tfrac{3}{2}\right)(0.50\text{ mol})(8.31\text{ J/mol K})(436.4\text{ K}) = 2700\text{ J}$$

$$E_{oxygen\,f} = n_{oxygen}\left(\tfrac{5}{2}RT_f\right) = \left(\tfrac{5}{2}\right)(0.25\text{ mol})(8.31\text{ J/mol K})(436.4\text{ K}) = 2300\text{ J}$$

(c) The change in the thermal energies are

$$E_{helium\,f} - E_{helium\,i} = 2720\text{ J} - 1870\text{ J} = 850\text{ J} \qquad E_{oxygen\,f} - E_{oxygen\,i} = 2266\text{ J} - 3116\text{ J} = -850\text{ J}$$

The helium gains energy and the oxygen loses energy.
(d) The final temperature can also be calculated as follows:

$$E_{helium\,f} = (n_{helium})\tfrac{3}{2}RT_f \Rightarrow 2720\text{ J} = (0.50\text{ mol})(1.5)(8.31\text{ J/mol K})T_f \Rightarrow T_f = 436.4\text{ K} \approx 436\text{ K}$$

18.53. Solve: Assuming that the systems are thermally isolated except from each other, the total energy for the two thermally interacting systems must remain the same. That is,

$$E_{1i} + E_{2i} = E_{1f} + E_{2f} \Rightarrow E_{1f} - E_{1i} = E_{2i} - E_{2f} = -(E_{2f} - E_{1f}) \Rightarrow \Delta E_1 = -\Delta E_2$$

No work is done on either system, so from the first law $\Delta E = Q$. Thus

$$Q_1 = -Q_2$$

That is, the heat lost by one system is gained by the other system.

18.55. Model: Isobaric processes occur at a constant temperature.
Visualize:

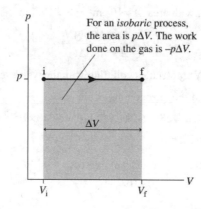

Solve:
(a) From the ideal gas law, if the volume quadruples at constant pressure then the temperature quadruples.

$$\frac{v'_{rms}}{v_{rms}} = \frac{\sqrt{\frac{3k_B T'}{m}}}{\sqrt{\frac{3k_B T}{m}}} = \frac{\sqrt{4T}}{\sqrt{T}} = 2$$

So v_{rms} increases by a factor of 2.
(b)

$$\lambda = \frac{1}{4\sqrt{2}\pi (N/V) r^2} = \frac{1}{4\sqrt{2}\pi \left(\frac{p}{k_B T}\right) r^2} = \frac{k_B T}{4\sqrt{2}\pi p r^2}$$

After the expansion everything in the expression for λ is the same except T which has quadrupled, so λ quadruples too.
(c)

$$\frac{E'_{th}}{E_{th}} = \frac{\frac{5}{2}nRT'}{\frac{5}{2}nRT} = \frac{4T}{T} = 4$$

So the thermal energy increases by a factor of 4.
(d) The molar specific heat at constant volume is $C_V = \frac{5}{2}R$ which is a constant. It does not depend on V, p, or T.

18.59. Solve: (a) The escape speed is the speed with which a mass m can leave the earth's surface and escape to infinity ($r_f = \infty$) with no left over speed ($v_f = 0$). The conservation of energy equation $K_f + U_f = K_i + U_i$ is

$$0 + 0 = \frac{1}{2}mv_{esc}^2 - \frac{GM_e m}{R_e} \Rightarrow v_{esc} = \sqrt{\frac{2GM_e}{R_e}}$$

The rms speed of a gas molecule is $v_{rms} = (3k_B T/m)^{1/2}$. Equating v_{esc} and v_{rms}, and squaring both sides, the temperature at which the rms speed equals the escape speed is

$$\frac{3k_B T}{m} = \frac{2GM_e}{R_e} \Rightarrow T = m\left(\frac{2GM_e}{3k_B R_e}\right)$$

For a nitrogen molecule, with $m = 28$ u, the temperature is

$$T = (28 \times 1.661 \times 10^{-27} \text{ kg}) \left(\frac{2(6.67 \times 10^{-11} \text{ N m}^2/\text{kg}^2)(5.98 \times 10^{24} \text{ kg})}{3(1.38 \times 10^{-23} \text{ J/K})(6.37 \times 10^6 \text{ m})} \right) = 141,000 \text{ K}$$

(b) For a hydrogen molecule, with $m = 2$ u, the temperature is less by a factor of 14, or $T = 10,100$ K.

(c) The average translational kinetic energy of a molecule is $\epsilon_{avg} = \frac{3}{2} k_B T = 6.1 \times 10^{-21}$ J at a typical atmosphere temperature of 20°C. The kinetic energy needed to escape is $K_{esc} = \frac{1}{2} m v_{esc}^2$. For nitrogen molecules, $K_{esc} = 2.9 \times 10^{-18}$ J. Thus $\epsilon_{avg}/K_{esc} = 0.002 = 0.2\%$. Earth will retain nitrogen in its atmosphere because the molecules are moving too slowly to escape. But for hydrogen molecules, with $K_{esc} = 2.1 \times 10^{-19}$ J, the ratio is $\epsilon_{avg}/K_{esc} = 0.03 = 3\%$. Thus a large enough fraction of hydrogen molecules are moving at escape speed, or faster, to allow hydrogen to leak out of the atmosphere into space. Consequently, earth's atmosphere does not contain hydrogen.

18.61. Solve: (a) The thermal energy is

$$E_{th} = (E_{th})_{N_2} + (E_{th})_{O_2} = \tfrac{5}{2} N_{N_2} k_B T + \tfrac{5}{2} N_{O_2} k_B T = \tfrac{5}{2} N_{total} k_B T$$

where N_{total} is the total number of molecules. The identity of the molecules makes no difference since both are diatomic. The number of molecules in the room is

$$N_{total} = \frac{pV}{k_B T} = \frac{(101,300 \text{ Pa})(2 \text{ m} \times 2 \text{ m} \times 2 \text{ m})}{(1.38 \times 10^{-23} \text{ J/K})(273 \text{ K})} = 2.15 \times 10^{26}$$

The thermal energy is

$$E_{th} = \tfrac{5}{2} (2.15 \times 10^{26})(1.38 \times 10^{-23} \text{ J/K})(273 \text{ K}) = 2.03 \times 10^6 \text{ J} \approx 2.0 \times 10^6 \text{ J}$$

(b) A 1 kg ball at height $y = 1$ m has a potential energy $U = mgy = 9.8$ J. The ball would need 9.8 J of initial kinetic energy to reach this height. The fraction of thermal energy that would have to be conveyed to the ball is

$$\frac{9.8 \text{ J}}{2.03 \times 10^6 \text{ J}} = 4.8 \times 10^{-6}$$

(c) A temperature change ΔT corresponds to a thermal energy change $\Delta E_{th} = \tfrac{5}{2} N_{total} k_B \Delta T$. But $\tfrac{5}{2} N_{total} k_B = E_{th}/T$. Using this, we can write

$$\Delta E_{th} = \frac{E_{th}}{T} \Delta T \Rightarrow \Delta T = \frac{\Delta E_{th}}{E_{th}} T = \frac{-9.8 \text{ J}}{2.03 \times 10^6 \text{ J}} 273 \text{ K} = -0.0013 \text{ K}$$

The room temperature would decrease by 0.0013 K or 0.0013°C.
(d) The situation with the ball at rest on the floor and in thermal equilibrium with the air is a very probable distribution of energy and thus a state with high entropy. Although energy would be conserved by removing energy from the air and transferring it to the ball, this would be a very *improbable* distribution of energy and thus a state of low entropy. The ball will not be spontaneously launched from the ground because this would require a decrease in entropy, in violation of the second law of thermodynamics.

As another way of thinking about the situation, the ball and the air are initially at the same temperature. Once even the slightest amount of energy is transferred from the air to the ball, the air's temperature will be less than that of the ball. Any further flow of energy from the air to the ball would be a situation in which heat energy is flowing from a colder object to a hotter object. This cannot happen because it would violate the second law of thermodynamics.

HEAT ENGINES AND REFRIGERATORS

Exercises and Problems

Section 19.1 Turning Heat into Work

Section 19.2 Heat Engines and Refrigerators

19.1. Solve: (a) The engine has a thermal efficiency of $\eta = 40\% = 0.40$ and a work output of 100 J per cycle. The heat input is calculated as follows:

$$\eta = \frac{W_{out}}{Q_H} \quad \Rightarrow \quad 0.40 = \frac{100 \text{ J}}{Q_H} \quad \Rightarrow \quad Q_H = 250 \text{ J}$$

(b) Because $W_{out} = Q_H - Q_C$, the heat exhausted is

$$Q_C = Q_H - W_{out} = 250 \text{ J} - 100 \text{ J} = 150 \text{ J}$$

19.5. Solve: (a) The heat extracted from the cold reservoir is calculated as follows:

$$K = \frac{Q_C}{W_{in}} \quad \Rightarrow \quad 4.0 = \frac{Q_C}{50 \text{ J}} \quad \Rightarrow \quad Q_C = 200 \text{ J}$$

(b) The heat exhausted to the hot reservoir is

$$Q_H = Q_C + W_{in} = 200 \text{ J} + 50 \text{ J} = 250 \text{ J}$$

Section 19.3 Ideal-Gas Heat Engines

Section 19.4 Ideal-Gas Refrigerators

19.9. Model: Process A is isochoric, process B is isothermal, process C is adiabatic, and process D is isobaric.
Solve: Process A is isochoric, so the increase in pressure increases the temperature and hence the thermal energy. Because $\Delta E_{th} = Q - W_s$ and $W_s = 0$ J, Q increases for process A. Process B is adiabatic, so $Q = 0$ J. W_s is positive because of the increase in volume. Since $Q = 0$ J $= W_s + \Delta E_{th}$, ΔE_{th} is negative for process B. Process C is isothermal, so T is constant and hence $\Delta E_{th} = 0$ J. The work done W_s is positive because the gas expands. Because $Q = W_s + \Delta E_{th}$, Q is positive for process B. Process D is isobaric, so the decrease in volume leads to a decrease in temperature and hence a decrease in the thermal energy. Due to the decrease in volume, W_s is negative. Because $Q = W_s + \Delta E_{th}$, Q also decreases for process D.

	ΔE_{th}	W_S	Q
A	+	0	+
B	−	+	0
C	0	+	+
D	−	−	−

19.11. Solve: The work done by the gas per cycle is the area inside the closed p-versus-V curve. The area inside the triangle is

$$W_{out} = \tfrac{1}{2}(3 \text{ atm} - 1 \text{ atm})(600 \times 10^{-6} \text{ m}^3 - 200 \times 10^{-6} \text{ m}^3)$$

$$= \tfrac{1}{2}\left(2 \text{ atm} \times \frac{1.013 \times 10^5 \text{ Pa}}{1 \text{ atm}}\right)(400 \times 10^{-6} \text{ m}^3) = 40 \text{ J}$$

19.15. Model: The heat engine follows a closed cycle.
Solve: The work done by the gas per cycle is the area inside the closed p-versus-V curve. We get

$$W_{out} = \tfrac{1}{2}(300 \text{ kPa} - 100 \text{ kPa})(600 \text{ cm}^3 - 300 \text{ cm}^3) = \tfrac{1}{2}(200 \times 10^3 \text{ Pa})(300 \times 10^{-6} \text{ m}^3) = 30 \text{ J}$$

Because $W_{out} = Q_H - Q_C$, the heat exhausted is

$$Q_C = Q_H - W_{out} = (225 \text{ J} + 90 \text{ J}) - 30 \text{ J} = 315 \text{ J} - 30 \text{ J} = 285 \text{ J}$$

19.17. Model: The Brayton cycle involves two adiabatic processes and two isobaric processes. The adiabatic processes involve compression and expansion through the turbine.
Solve: The thermal efficiency for the Brayton cycle is $\eta_B = 1 - r_p^{(1-\gamma)/\gamma}$, where $\gamma = C_P/C_V$ and r_p is the pressure ratio. For a diatomic gas $\gamma = 1.4$. For an adiabatic process,

$$p_1 V_1^\gamma = p_2 V_2^\gamma \quad \Rightarrow \quad p_2/p_1 = (V_1/V_2)^\gamma$$

Because the volume is halved, $V_2 = \tfrac{1}{2}V_1$ so

$$r_p = p_2/p_1 = (2)^\gamma = 2^{1.4} = 2.639$$

The efficiency is

$$\eta_B = 1 - (2.639)^{-0.4/1.4} = 0.24$$

Section 19.5 The Limits of Efficiency

Section 19.6 The Carnot Cycle

19.19. Model: The efficiency of a Carnot engine (η_{Carnot}) depends only on the temperatures of the hot and cold reservoirs. On the other hand, the thermal efficiency (η) of a heat engine depends on the heats Q_H and Q_C.
Solve: (a) According to the first law of thermodynamics, $Q_H = W_{out} + Q_C$. For engine (a), $Q_H = 500 \text{ J}$, $Q_C = 200 \text{ J}$ and $W_{out} = 300 \text{ J}$, so the first law of thermodynamics is obeyed. For engine (b), $Q_H = 500 \text{ J}, Q_C = 200 \text{ J}$ and $W_{out} = 200 \text{ J}$, so the first law is violated. For engine (c) $Q_H = 300 \text{ J}, Q_C = 200 \text{ J}$ and $W_{out} = 100 \text{ J}$, so the first law of thermodynamics is obeyed.
(b) For the three heat engines, the maximum or Carnot efficiency is

$$\eta_{Carnot} = 1 - \frac{T_C}{T_H} = 1 - \frac{300 \text{ K}}{600 \text{ K}} = 0.50$$

Engine (a) has

$$\eta = 1 - \frac{Q_C}{Q_H} = \frac{W_{out}}{Q_H} = \frac{300 \text{ J}}{500 \text{ J}} = 0.60$$

This is larger than η_{Carnot}, thus violating the second law of thermodynamics. For engine (b),

$$\eta = \frac{W_{out}}{Q_H} = \frac{200 \text{ J}}{500 \text{ J}} = 0.40 < \eta_{Carnot}$$

so the second law is obeyed. Engine (c) has a thermal efficiency of

$$\eta = \frac{100 \text{ J}}{300 \text{ J}} = 0.33 < \eta_{Carnot}$$

so the second law of thermodynamics is obeyed.

19.21. Model: The efficiency of a Carnot engine depends only on the absolute temperatures of the hot and cold reservoirs.
Solve: The efficiency of a Carnot engine is

$$\eta_{Carnot} = 1 - \frac{T_C}{T_H} \quad \Rightarrow \quad 0.60 = 1 - \frac{T_C}{(427 + 273) \text{ K}} \quad \Rightarrow \quad T_C = 280 \text{ K} = 7°C$$

Assess: A "real" engine would need a lower temperature than 7°C to provide 60% efficiency because no real engine can match the Carnot efficiency.

19.25. Model: We will use Equation 19.27 for the efficiency of a Carnot engine.

$$\eta_{Carnot} = 1 - \frac{T_C}{T_H}$$

We are given $T_H = 673$ K and the original efficiency $\eta_{Carnot} = 0.40$.
Solve: First solve for T_C.

$$T_C = T_H(1 - \eta_{Carnot}) = (673 \text{ K})(1 - 0.40) = 404 \text{ K}$$

Solve for T_C' again with $\eta'_{Carnot} = 0.60$.

$$T_C' = T_H(1 - \eta'_{Carnot}) = (673 \text{ K})(1 - 0.60) = 269 \text{ K}$$

The difference of these T_C values is 135 K, so the temperature of the cold reservoir should be decreased by 135°C to raise the efficiency from 40% to 60%.
Assess: We expected to have to lower T_C by quite a bit to get the better efficiency.

19.29. Model: The minimum possible value of T_C occurs with a Carnot refrigerator.
Solve: (a) For the refrigerator, the coefficient of performance is

$$K = \frac{Q_C}{W_{in}} \quad \Rightarrow \quad Q_C = KW_{in} = (5.0)(10 \text{ J}) = 50 \text{ J}$$

The heat energy exhausted per cycle is

$$Q_H = Q_C + W_{in} = 50 \text{ J} + 10 \text{ J} = 60 \text{ J}$$

(b) If the hot-reservoir temperature is 27°C − 300 K, the lowest possible temperature of the cold reservoir can be obtained as follows:

$$K_{Carnot} = \frac{T_C}{T_H - T_C} \quad \Rightarrow \quad 5.0 = \frac{T_C}{300 \text{ K} - T_C} \quad \Rightarrow \quad T_C = 250 \text{ K} = -23°C$$

19.33. Solve: An adiabatic process has $Q = 0$ and thus, from the first law, $W_s = -\Delta E_{th}$. For any ideal-gas process, $\Delta E_{th} = nC_V \Delta T$, so $W_s = -nC_V \Delta T$. We can use the ideal-gas law to find

$$T = \frac{pV}{nR} \quad \Rightarrow \quad \Delta T = \frac{\Delta(pV)}{nR} = \frac{(pV)_f - (pV)_i}{nR} = \frac{p_f V_f - p_i V_i}{nR}$$

Consequently, the work is

$$W_s = -nC_V \Delta T = -nC_V \left(\frac{p_f V_f - p_i V_i}{nR} \right) = -\frac{C_V}{R}(p_f V_f - p_i V_i)$$

Because $C_P = C_V + R$, we can use the specific heat ratio γ to find

$$\gamma = \frac{C_P}{C_V} = \frac{C_V + R}{C_V} = \frac{C_V/R + 1}{C_V/R} \quad \Rightarrow \quad \frac{C_V}{R} = \frac{1}{\gamma - 1}$$

With this, the work done in an adiabatic process is

$$W_s = -\frac{C_V}{R}(p_f V_f - p_i V_i) = -\frac{1}{\gamma - 1}(p_f V_f - p_i V_i) = \frac{p_f V_f - p_i V_i}{1 - \gamma}$$

19.35. Solve: For any heat engine, $\eta = 1 - Q_C/Q_H$. For a Carnot heat engine, $\eta_{\text{Carnot}} = 1 - T_C/T_H$. Thus a property of the Carnot cycle is that $Q_C/Q_H = T_C/T_H$. Consequently, the coefficient of performance of a Carnot refrigerator is

$$K_{\text{Carnot}} = \frac{Q_C}{W_{\text{in}}} = \frac{Q_C}{Q_H - Q_C} = \frac{Q_C/Q_H}{1 - Q_C/Q_H} = \frac{T_C/T_H}{1 - T_C/T_H} = \frac{T_C}{T_H - T_C}$$

19.37. Model: We will use the Carnot engine to find the maximum possible efficiency of a floating power plant.
Solve: The efficiency of a Carnot engine is

$$\eta_{\text{max}} = \eta_{\text{Carnot}} = 1 - \frac{T_C}{T_H} = 1 - \frac{(273 + 5)\ \text{K}}{(273 + 30)\ \text{K}} = 0.0825 \approx 8.3\%$$

19.41. Model: Assume the soda is essentially made of water. We are given $T_H = 328$ K, $T_C = 253$ K, and $Q_H = 250$ J. See Figure 19.11.
Solve: The total amount of heat to transfer from the soda is

$$(Q_C)_{\text{Total}} = M c \Delta T = V \rho c \Delta T = (5.00 \times 10^{-4}\ \text{m}^3)(1000\ \text{kg/m}^3)(4190\ \text{J/kg K})(20\ \text{K}) = 41,900\ \text{J}$$

For a Carnot cycle $\eta_{\text{Carnot}} = 1 - \dfrac{T_C}{T_H}$ but also $\eta = 1 - \dfrac{Q_C}{Q_H}$, so $\dfrac{Q_C}{Q_H} = \dfrac{T_C}{T_H}$. Therefore, the heat extracted from the soda each cycle is

$$Q_C = Q_H \frac{T_C}{T_H} = (250\ \text{J})\left(\frac{253\ \text{K}}{328\ \text{K}}\right) = 192.8\ \text{J}$$

To remove 41,900 J will therefore take

$$\frac{(Q_C)_{\text{Total}}}{Q_C} = \frac{41,900\ \text{J}}{192.8\ \text{J}} = 218\ \text{cycles}$$

19.45. Visualize: We are given $T_C = 275$ K, $T_H = 295$ K. We are also given that in one second $W_{\text{in}} = 100$ J and $Q_C = (1\ \text{s})(100\ \text{kJ/min})(1\ \text{min}/60\ \text{s}) = 1667$ J.
Solve: The coefficient of performance of a refrigerator is given in Equation 19.8.

$$K = \frac{Q_C}{W_{\text{in}}} = \frac{1667\ \text{J}}{100\ \text{J}} = 16.67$$

However the coefficient of performance of a Carnot refrigerator is given in Equation 19.28.

$$K_{\text{Carnot}} = \frac{T_C}{T_H - T_C} = \frac{275\ \text{K}}{20\ \text{K}} = 13.75$$

However, informal statement #8 of the second law says that the coefficient of performance cannot exceed the Carnot coefficient of performance, so the salesman is making false claims. You should not buy the DreamFridge.
Assess: The second law imposes real-world restrictions.

19.47. Model: The power plant is to be treated as a heat engine.
Solve: (a) Every hour 300 metric tons or 3×10^5 kg of coal is burnt. The volume of coal is

$$\left(\frac{3 \times 10^5\ \text{kg}}{1\ \text{h}}\right)\left(\frac{\text{m}^3}{1500\ \text{kg}}\right)(24\ \text{h}) = 4800\ \text{m}^3$$

The height of the room will be 48 m.

(b) The thermal efficiency of the power plant is

$$\eta = \frac{W_{out}}{Q_H} = \frac{7.50 \times 10^8 \text{ J/s}}{\left(\dfrac{3 \times 10^5 \text{ kg}}{1 \text{ h}}\right)\left(\dfrac{28 \times 10^6 \text{ J}}{\text{kg}}\right)\left(\dfrac{1 \text{ h}}{3600 \text{ s}}\right)} = \frac{7.50 \times 10^8 \text{ J}}{2.333 \times 10^9 \text{ J}} = 0.32 = 32\%$$

Assess: An efficiency of 32% is typical of power plants.

19.51. Visualize: If we do this problem on a "per-second" basis then in one second $Q_C = (1 \text{ s})(5.0 \times 10^5 \text{ J/min})$

$(1 \text{ min}/60 \text{ s}) = 8.33 \times 10^3$ J. $Q_H = (1 \text{ s})(8.0 \times 10^5 \text{ J/min})(1 \text{ min}/60 \text{ s}) = 13.33 \times 10^3$ J.

Solve: (a) Again, in one second

$$W_{in} = Q_H - Q_C = 13.33 \times 10^3 \text{ J} - 8.33 \times 10^3 \text{ J} = 5.0 \times 10^3 \text{ J}$$

Since this is per second, the power required by the compressor is $P = 5.0$ kW.

(b) The coefficient of performance is

$$K = \frac{Q_C}{W_{in}} = \frac{8.33 \times 10^3 \text{ J}}{5.0 \times 10^3 \text{ J}} = 1.7$$

Assess: The result is typical for air conditioners.

19.53. Model: The heat engine follows a closed cycle. For a diatomic gas, $C_V = \frac{5}{2}R$ and $C_P = \frac{7}{2}R$.

Visualize: Please refer to Figure P19.53.

Solve: (a) Since $T_1 = 293$ K, the number of moles of the gas is

$$n = \frac{p_1 V_1}{R T_1} = \frac{(0.5 \times 1.013 \times 10^5 \text{ Pa})(10 \times 10^{-6} \text{ m}^3)}{(8.31 \text{ J/mol K})(293 \text{ K})} = 2.08 \times 10^{-4} \text{ mol}$$

At point 2, $V_2 = 4V_1$ and $p_2 = 3p_1$. The temperature is calculated as follows:

$$\frac{p_1 V_1}{T_1} = \frac{p_2 V_2}{T_2} \implies T_2 = \frac{p_2}{p_1}\frac{V_2}{V_1}T_1 = (3)(4)(293 \text{ K}) = 3516 \text{ K}$$

At point 3, $V_3 = V_2 = 4V_1$ and $p_3 = p_1$. The temperature is calculated as before:

$$T_3 = \frac{p_3}{p_1}\frac{V_3}{V_1}T_1 = (1)(4)(293 \text{ K}) = 1172 \text{ K}$$

For process $1 \rightarrow 2$, the work done is the area under the p-versus-V curve. That is,

$$W_s = (0.5 \text{ atm})(40 \text{ cm}^3 - 10 \text{ cm}^3) + \frac{1}{2}(1.5 \text{ atm} - 0.5 \text{ atm})(40 \text{ cm}^3 - 10 \text{ cm}^3)$$

$$= (30 \times 10^{-6} \text{ m}^3)(1 \text{ atm})\left(\frac{1.013 \times 10^5 \text{ Pa}}{1 \text{ atm}}\right) = 3.04 \text{ J}$$

The change in the thermal energy is

$$\Delta E_{th} = n C_V \Delta T = (2.08 \times 10^{-4} \text{ mol})\tfrac{5}{2}(8.31 \text{ J/mol K})(3516 \text{ K} - 293 \text{ K}) = 13.93 \text{ J}$$

The heat is $Q = W_s + \Delta E_{th} = 16.97$ J. For process $2 \rightarrow 3$, the work done is $W_s = 0$ J and

$$Q = \Delta E_{th} = n C_V \Delta T = n\left(\tfrac{5}{2}R\right)(T_3 - T_2)$$

$$= (2.08 \times 10^{-4} \text{ mol})\tfrac{5}{2}(8.31 \text{ J/mol K})(1172 \text{ K} - 3516 \text{ K}) = -10.13 \text{ J}$$

For process $3 \rightarrow 1$,

$$W_s = (0.5 \text{ atm})(10 \text{ cm}^3 - 40 \text{ cm}^3) = (0.5 \times 1.013 \times 10^5 \text{ Pa})(-30 \times 10^{-6} \text{ m}^3) = -1.52 \text{ J}$$

$$\Delta E_{th} = n C_V \Delta T = (2.08 \times 10^{-4} \text{ mol})\tfrac{5}{2}(8.31 \text{ J/mol K})(293 \text{ K} - 1172 \text{ K}) = -3.80 \text{ J}$$

The heat is $Q = \Delta E_{th} + W_s = -5.32$ J.

	W_s (J)	Q (J)	ΔE_{th}
$1 \rightarrow 2$	3.04	16.97	13.93
$2 \rightarrow 3$	0	-10.13	-10.13
$3 \rightarrow 1$	-1.52	-5.32	-3.80
Net	1.52	1.52	0

(b) The efficiency of the engine is

$$\eta = \frac{W_{net}}{Q_H} = \frac{1.52 \text{ J}}{16.97 \text{ J}} = 0.090 = 9.0\%$$

(c) The power output of the engine is

$$500 \left(\frac{\text{revolutions}}{\text{min}} \right) \left(\frac{1 \text{ min}}{60 \text{ s}} \right) \left(\frac{W_{net}}{\text{revolution}} \right) = \left(\frac{500}{60} \right)(1.52 \text{ J/s}) = 13 \text{ W}$$

Assess: Note that more than two significant figures are retained in part (a) because the results are intermediate. For a closed cycle, as expected, $(W_s)_{net} = Q_{net}$ and $(\Delta E_{th})_{net} = 0$ J.

19.57. Model: For the closed cycle of the heat engine, process $1 \rightarrow 2$ is adiabatic, process $2 \rightarrow 3$ is isothermal, and process $3 \rightarrow 1$ is isobaric. For a diatomic gas $C_V = \frac{5}{2}R$ and $\gamma = \frac{7}{5}$.

Solve: (a) From the graph $p_1 = 100$ kPa.
The number of moles of gas is

$$n = \frac{p_2 V_2}{RT_2} = \frac{(4.00 \times 10^5 \text{ Pa})(1000 \times 10^{-6} \text{ m}^3)}{(8.31 \text{ J/mol K})(400 \text{ K})} = 0.1203 \text{ mol}$$

Using $p_1 V_1^\gamma = p_2 V_2^\gamma$, and reading $V_2 = 1000$ cm³ from the graph,

$$V_1 = V_2 \left(\frac{p_2}{p_1} \right)^{1/\gamma} = (1000 \text{ cm}^3) \left(\frac{4.00 \times 10^5 \text{ Pa}}{1.00 \times 10^5 \text{ Pa}} \right)^{5/7} = 2692 \text{ cm}^3 \approx 2690 \text{ cm}^3$$

With $p_1, V_1,$ and n determined, we can find T_1 using the ideal-gas equation:

$$T_1 = \frac{p_1 V_1}{nR} = \frac{(1.00 \times 10^5 \text{ Pa})(2692 \times 10^{-6} \text{ m}^3)}{(8.31 \text{ J/mol K})(0.1203 \text{ mol})} = 269 \text{ K}$$

(b) For the adiabatic process $1 \rightarrow 2$, $Q = 0$ J and

$$W_s = \frac{p_2 V_2 - p_1 V_1}{1 - \gamma} = \frac{(4.00 \times 10^5 \text{ Pa})(1.000 \times 10^{-3} \text{ m}^3) - (1.00 \times 10^5 \text{ Pa})(2692 \times 10^{-6} \text{ m}^3)}{1 - 1.4} = -327 \text{ J}$$

Because $\Delta E_{th} = -W_s + Q$, $\Delta E_{th} = -W_s = 327$ J. For the isothermal process $2 \rightarrow 3$, $\Delta E_{th} = 0$ J. From Table 19.1,

$$W_s = nRT_2 \ln \frac{V_3}{V_2} = (0.1203 \text{ mol})(8.31 \text{ J/mol K})(400 \text{ K}) \ln \left(\frac{4000 \text{ cm}^3}{1000 \text{ cm}^3} \right) = 555 \text{ J}$$

From the first law of thermodynamics, $Q = W_s = 555$ J for process $2 \rightarrow 3$. For the isobaric process $3 \rightarrow 1$, $W_s \rightarrow 0$ J and

$$Q = \Delta E_{th} = nC_P \Delta T = n \left(\frac{7}{2} R \right)(T_1 - T_3) = (0.1203 \text{ mol}) \left(\frac{7}{2} \right)(8.31 \text{ J/mol K})(269 \text{ K} - 400 \text{ K}) = -458 \text{ J}$$

	ΔE_{th} (J)	W_s (J)	Q (J)
$1 \rightarrow 2$	327	-327	0
$2 \rightarrow 3$	0	555	555
$3 \rightarrow 1$	-458	0	-458
Net	-131	228	97

(c) The work per cycle is 97 J and the thermal efficiency of the engine is

$$\eta = \frac{W_s}{Q_H} = \frac{97\ \text{J}}{555\ \text{J}} = 0.17 = 17\%$$

Assess: This efficiency is less than the Carnot efficiency $\eta_{\text{Carnot}} = 1 - T_C/T_H = 1 - 269/400 = 33\%$ and so is possible.

19.59. Model: Process $1 \rightarrow 2$ of the cycle is isochoric, process $2 \rightarrow 3$ is isothermal, and process $3 \rightarrow 1$ is isobaric. For a monatomic gas, $C_V = \frac{3}{2}R$ and $C_P = \frac{5}{2}R$.

Visualize: Please refer to Figure P19.59.

Solve: (a) At point 1: The pressure $p_1 = 1\ \text{atm} = 1.013 \times 10^5\ \text{Pa}$ and the volume $V_1 = 1000 \times 10^{-6}\ \text{m}^3 = 1 \times 10^{-3}\ \text{m}^3$. The number of moles is

$$n = \frac{0.120\ \text{g}}{4\ \text{g/mol}} = 0.03\ \text{mol}$$

Using the ideal-gas law,

$$T_1 = \frac{p_1 V_1}{nR} = \frac{(1.013 \times 10^5\ \text{Pa})(1.0 \times 10^{-3}\ \text{m}^3)}{(0.030\ \text{mol})(8.31\ \text{J/mol K})} = 406\ \text{K} \approx 0.4\ \text{kK}$$

At point 2: The pressure $p_2 = 5\ \text{atm} = 5.06 \times 10^5\ \text{Pa}$ and $V_2 = 1 \times 10^{-3}\ \text{m}^3$. The temperature is

$$T_2 = \frac{p_2 V_2}{nR} = \frac{(5.06 \times 10^5\ \text{Pa})(1.0 \times 10^{-3}\ \text{m}^3)}{(0.030\ \text{mol})(8.31\ \text{J/mol K})} = 2030\ \text{K} \approx 2\ \text{kK}$$

At point 3: The pressure is $p_3 = 1\ \text{atm} = 1.013 \times 10^5\ \text{Pa}$ and the temperature is $T_3 = T_2 = 2030\ \text{K}$. The volume is

$$V_3 = V_2 \frac{p_2}{p_3} = (1 \times 10^{-3}\ \text{m}^3)\left(\frac{5\ \text{atm}}{1\ \text{atm}}\right) = 5 \times 10^{-3}\ \text{m}^3$$

(b) For isochoric process $1 \rightarrow 2$, $W_{1 \rightarrow 2} = 0\ \text{J}$ and

$$Q_{1 \rightarrow 2} = nC_V \Delta T = (0.030\ \text{mol})\left(\tfrac{3}{2}R\right)(2030\ \text{K} - 406\ \text{K}) = 607\ \text{J}$$

For isothermal process $2 \rightarrow 3$, $\Delta E_{\text{th}\ 2 \rightarrow 3} = 0\ \text{J}$ and

$$Q_{2 \rightarrow 3} = W_{2 \rightarrow 3} = nRT_2 \ln\frac{V_3}{V_2} = (0.030\ \text{mol})(8.31\ \text{J/mol K})(2030\ \text{K})\ln\left(\frac{5.0 \times 10^{-3}\ \text{m}^3}{1.0 \times 10^{-3}\ \text{m}^3}\right) = 815\ \text{J}$$

For isobaric process $3 \rightarrow 1$,

$$W_{3 \rightarrow 1} = p_3 \Delta V = (1.013 \times 10^5\ \text{Pa})(1.0 \times 10^{-3}\ \text{m}^3 - 5.0 \times 10^{-3}\ \text{m}^3) = -405\ \text{J}$$

$$Q_{3 \rightarrow 1} = nC_P \Delta T = (0.030\ \text{mol})\left(\tfrac{5}{2}\right)(8.31\ \text{J/mol K})(406\ \text{K} - 2030\ \text{K}) = -1012\ \text{J}$$

The total work done is $W_{\text{net}} = W_{1 \rightarrow 2} + W_{2 \rightarrow 3} + W_{3 \rightarrow 1} = 410\ \text{J}$. The total heat input is $Q_H = Q_{1 \rightarrow 2} + Q_{2 \rightarrow 3} = 1422\ \text{J}$. The thermal efficiency of the engine is

$$\eta = \frac{W_{\text{net}}}{Q_H} = \frac{410\ \text{J}}{1422\ \text{J}} = 20\%$$

(c) The maximum possible efficiency of a heat engine that operates between T_{max} and T_{min} is

$$\eta_{\text{max}} = 1 - \frac{T_{\text{min}}}{T_{\text{max}}} = 1 - \frac{406\ \text{K}}{2030\ \text{K}} = 80\%$$

Assess: The actual efficiency of an engine is less than the maximum possible efficiency.

19.61. Model: The closed cycle in this heat engine includes adiabatic process $1 \rightarrow 2$, isobaric process $2 \rightarrow 3$, and isochoric process $3 \rightarrow 1$. For a diatomic gas, $C_V = \frac{5}{2}R$, $C_P = \frac{7}{2}R$, and $\gamma = \frac{7}{5} = 1.4$.

Visualize: Please refer to Figure P19.61.

Solve: (a) We can find the temperature T_2 from the ideal-gas equation as follows:

$$T_2 = \frac{p_2 V_2}{nR} = \frac{(4.0 \times 10^5 \text{ Pa})(1.0 \times 10^{-3} \text{ m}^3)}{(0.020 \text{ mol})(8.31 \text{ J/mol K})} = 2407 \text{ K} \approx 2.4 \text{ kK}$$

We can use the equation $p_2 V_2^\gamma = p_1 V_1^\gamma$ to find V_1,

$$V_1 = V_2 \left(\frac{p_2}{p_1} \right)^{1/\gamma} = (1.0 \times 10^{-3} \text{ m}^3) \left(\frac{4.0 \times 10^5 \text{ Pa}}{1.0 \times 10^5 \text{ Pa}} \right)^{1/1.4} = 2.692 \times 10^{-3} \text{ m}^3$$

The ideal-gas equation can now be used to find T_1,

$$T_1 = \frac{p_1 V_1}{nR} = \frac{(1.0 \times 10^5 \text{ Pa})(2.692 \times 10^{-3} \text{ m}^3)}{(0.020 \text{ mol})(8.31 \text{ J/mol K})} = 1620 \text{ K} \approx 1.6 \text{ kK}$$

At point 3, $V_3 = V_1$ so we have

$$T_3 = \frac{p_3 V_3}{nR} = \frac{(4 \times 10^5 \text{ Pa})(2.692 \times 10^{-3} \text{ m}^3)}{(0.020 \text{ mol})(8.31 \text{ J/mol K})} = 6479 \text{ K} \approx 6.5 \text{ kK}$$

(b) For adiabatic process $1 \rightarrow 2$, $Q = 0$ J, $\Delta E_{th} = -W_s$, and

$$W_s = \frac{p_2 V_2 - p_1 V_1}{1 - \gamma} = \frac{nR(T_2 - T_1)}{1 - \gamma} = \frac{(0.020 \text{ mol})(8.31 \text{ J/mol K})(2407 \text{ K} - 1620 \text{ K})}{(1 - 1.4)} = -327.0 \text{ J}$$

For isobaric process $2 \rightarrow 3$,

$$Q = nC_p \Delta T = n \left(\tfrac{7}{2} R \right)(\Delta T) = (0.020 \text{ mol}) \tfrac{7}{2} (8.31 \text{ J/mol K})(6479 \text{ K} - 2407 \text{ K}) = 2369 \text{ J}$$

$$\Delta E_{th} = nC_V \Delta T = n \left(\tfrac{5}{2} R \right) \Delta T = 1692 \text{ J}$$

The work done is the area under the p-versus-V graph. Hence,

$$W_s = (4.0 \times 10^5 \text{ Pa})(2.692 \times 10^{-3} \text{ m}^3 - 1.0 \times 10^{-3} \text{ m}^3) = 677 \text{ J}$$

For isochoric process $3 \rightarrow 1$, $W_s = 0$ J and

$$\Delta E_{th} = Q = nC_V \Delta T = (0.020 \text{ mol}) \left(\tfrac{5}{2} \right)(8.31 \text{ J/mol K})(1620 \text{ K} - 6479 \text{ K}) = -2019 \text{ J}$$

	ΔE_{th} (J)	W_S (J)	Q (J)
$1 \rightarrow 2$	327	−327	0
$2 \rightarrow 3$	1692	677	2369
$3 \rightarrow 1$	−2019	0	−2019
Net	0	350	350

(c) The engine's thermal efficiency is

$$\eta = \frac{W_{net}}{Q_H} = \frac{350 \text{ J}}{2369 \text{ J}} = 0.15 = 15\%$$